SpringerBriefs in Energy

SpringerBriefs in Energy presents concise summaries of cutting-edge research and practical applications in all aspects of Energy. Featuring compact volumes of 50 to 125 pages, the series covers a range of content from professional to academic. Typical topics might include:

- A snapshot of a hot or emerging topic
- A contextual literature review
- A timely report of state-of-the art analytical techniques
- An in-depth case study
- A presentation of core concepts that students must understand in order to make independent contributions.

Briefs allow authors to present their ideas and readers to absorb them with minimal time investment.

Briefs will be published as part of Springer's eBook collection, with millions of users worldwide. In addition, Briefs will be available for individual print and electronic purchase. Briefs are characterized by fast, global electronic dissemination, standard publishing contracts, easy-to-use manuscript preparation and formatting guidelines, and expedited production schedules. We aim for publication 8–12 weeks after acceptance.

Both solicited and unsolicited manuscripts are considered for publication in this series. Briefs can also arise from the scale up of a planned chapter. Instead of simply contributing to an edited volume, the author gets an authored book with the space necessary to provide more data, fundamentals and background on the subject, methodology, future outlook, etc.

SpringerBriefs in Energy contains a distinct subseries focusing on Energy Analysis and edited by Charles Hall, State University of New York. Books for this subseries will emphasize quantitative accounting of energy use and availability, including the potential and limitations of new technologies in terms of energy returned on energy invested.

More information about this series at http://www.springer.com/series/8903

Shin-ichi Nakao · Katsunori Yogo ·
Kazuya Goto · Teruhiko Kai ·
Hidetaka Yamada

Advanced CO$_2$ Capture Technologies

Absorption, Adsorption, and Membrane
Separation Methods

 Springer

Shin-ichi Nakao
Research Institute of Innovative
Technology for the Earth (RITE)
Kyoto, Japan

Katsunori Yogo
Research Institute of Innovative
Technology for the Earth (RITE)
Kyoto, Japan

Kazuya Goto
Research Institute of Innovative
Technology for the Earth (RITE)
Kyoto, Japan

Teruhiko Kai
Research Institute of Innovative
Technology for the Earth (RITE)
Kyoto, Japan

Hidetaka Yamada
Research Institute of Innovative
Technology for the Earth (RITE)
Kyoto, Japan

ISSN 2191-5520 ISSN 2191-5539 (electronic)
SpringerBriefs in Energy
ISBN 978-3-030-18857-3 ISBN 978-3-030-18858-0 (eBook)
https://doi.org/10.1007/978-3-030-18858-0

This Springer imprint is published by the registered company Springer Nature Switzerland AG.
The registered company address is: Gewerbestrasse 11, 6330 Cham, Switzerland

Contents

Chapter 1
Introduction

Abstract As an introduction, this chapter describes the importance of CO_2-capture technologies in the context of limiting the global temperature increase.

Keywords CCS · CO_2 capture · 2 °C scenario · IPCC

1.1 The Paris Agreement

The Paris Agreement, which is an agreement within the United Nations Framework Convention on Climate Change (UNFCCC), has adopted in December 2015 and entered into force in November 2016. Its central aim is to strengthen the global response to the threat of climate change by holding the increase in the global average temperature to well below 2 °C above pre-industrial levels and pursuing efforts to limit the temperature increase to 1.5 °C above pre-industrial levels.

The Intergovernmental Panel on Climate Change (IPCC), which takes a role of assessing the science related to climate change and providing a scientific basis, is a source of scientific information and technical guidance for Parties to the UNFCCC, its Kyoto Protocol, and Paris Agreement. Following the Paris Agreement, the special report, Global Warming of 1.5 °C, has been approved by the IPCC and released in October 2018 [1]. This report states:

> Human activities are estimated to have caused approximately 1.0 °C of global warming above pre-industrial levels, with a likely range of 0.8 °C to 1.2 °C. Global warming is likely to reach 1.5 °C between 2030 and 2052 if it continues to increase at the current rate.

1.2 CO_2 Capture and Storage

The above special report finds that limiting global warming to 1.5 °C is not impossible. However, it would require an urgent global transformation in energy and industrial systems. In particular, it reports that for energy pathways limiting global warming to 1.5–2 °C, in addition to renewable, which is the most contributing pathway, CO_2

capture and storage (CCS) is still an indispensable technology. However, there is uncertainty in the future deployment of CCS:

> Given the importance of CCS in most mitigation pathways and its current slow pace of improvement, the large-scale deployment of CCS as an option depends on the further development of the technology in the near term.

CCS involves the trapping of CO_2 from the emissions generated during fossil fuel combustion from such sources as electric power plants, steel-making plants, cement plants, and factories and includes the subsequent sequestration of the captured CO_2 in geological formations [2]. However, the costs associated with capturing CO_2 from emission sources account for the greater part of overall CCS expenditures. Therefore, it is important to reduce capture costs to allow the practical deployment of CCS. In this context, developing advanced CO_2-capture technologies is an urgent priority for the deployment of CCS.

1.3 CO_2-Capture Methods

Because the optimum CO_2-capture process differs from the viewpoint of emission scale, concentration, pressure, and so forth at CO_2 source point, a variety of capture methods are required. The mainstream of CO_2-capture technology contains three approaches: absorption method, adsorption method, and membrane separation. Following this chapter, the book briefly covers these technologies in Chaps. 3 (absorption), 4 (adsorption), and 5 (membrane), after describing the underlying basic chemistry in Chap. 2. The advanced CO_2-capture technologies being developed by the authors (RITE [3]) are also reported there. It serves as a valuable reference resource for researchers, teachers, and students interested in CO_2 problems, providing essential information on how to capture CO_2 from various types of gases efficiently. It is also of interest to practitioners and academics, as it discusses the performance of the latest technologies applied in large-scale emission sources.

References

1. IPCC (2018) Special report on global warming of 1.5 °C
2. IPCC (2005) Special report on carbon dioxide capture and storage
3. RITE (Research Institute of Innovative Technology for the Earth). www.rite.or.jp/en/

Chapter 2
Chemistry of Amine-Based CO_2 Capture

Abstract Amines are the most widely utilized chemicals for CO_2 capture in a variety of capture methods based on the reversible reactions between amines and CO_2 since their moderate interaction allows effective "catch and release." In this chapter, the chemistry related to the CO_2-capture technology is reviewed with a special focus on amine–CO_2 reactions. Physical and chemical properties of the amines used in CO_2 capture are described at the molecular level.

Keywords Amine · Chemical reaction · Reaction mechanism · Solvation effect · Substituent effect

2.1 Amines for CO_2 Capture

Amines are derivatives of ammonia that contain one or more nitrogen atoms. In nature, a lot of amine compounds appear in the biological activities of living organisms. Amino acids serve as the building blocks of proteins. "Vitamin" was named so, because of the discovery of a nitrogenous substance essential for life [17]. Deoxyribonucleic acid, or DNA, stores information as a code made up of four kinds of nitrogenous bases.

In industry, amines also play important roles in a variety of applications. The lower alkylamines derived from feedstocks containing C_2–C_5 carbon chains are used in pharmaceutical, agricultural, rubber chemical industries, while the C_1 alkylamines, i.e., methylamines are produced as intermediates for solvents such as N-methylpyrrolidone, dimethylformamide and dimethylacetamide, agricultural chemicals, surfactants, and water treatment chemicals. In most cases, all these commercial alkylamines are synthesized from the reaction of an alcohol with ammonia [12].

Amine scrubbing, the method for separating acidic gases with an aqueous amine, has been widely used in industry [18] since its basic process was patented in 1930 [3]. In this patent, you can find the following description:

> I have furthermore discovered that these properties are possessed only by these amines which have certain chemical characteristics as to arrangement of atoms and certain physical characteristics. The presence of oxygen in addition to nitrogen and hydrogen is not objectionable, but the oxygen must not be present in a carboxyl group or a carbonyl group, although it

© The Author(s), under exclusive license to Springer Nature Switzerland AG 2019   3
S. Nakao et al., *Advanced C O2 Capture Technologies*,
SpringerBriefs in Energy, https://doi.org/10.1007/978-3-030-18858-0_2

may be present in a hydroxyl group. The amine must be either solid or liquid at ordinary room temperature, and must have a boiling point not substantially below 100 °C. It must be soluble in water or other liquid which does not form a stable compound with the acidic gas or other gases associated therewith, and which has a boiling point not below the temperature of elective gas elimination.

This description explains the reason why monoethanolamine (MEA) and other some alkanolamines have been adopted for amine scrubbing. In addition, easy availability is also of advantage.

Since the end of the twentieth century, amine scrubbing has been in the spotlight as a technology for reducing CO_2 emission. It is notable that in the context of global warming countermeasures, the amount of CO_2 that should be reduced is too large to treat in sustainable cycles. One of the most promising solutions is CO_2 capture and storage (CCS). In the CCS, CO_2 is selectively separated from large-point sources such as fossil fuel power plants, transported to a storage site, and then stored in appropriate geological formations that are typically located a few kilometers below the earth's surface.

Although the amine scrubbing is the most mature technology for CO_2 separation, its application to CCS has posed economic and environmental challenges such as its energy penalty, amine degradation, and emission [18]. Therefore, the conventional amine scrubbing method has been restudied for improvements in the twenty-first century. Furthermore, alternative methods based on absorption, adsorption, and membrane separation have been extensively studied all over the world.

Still, amines have been the most widely utilized chemicals in various developing technologies for CO_2 capture [10] because the reversible reactions between amines and CO_2 allow effective "catch and release" through their moderate interaction. More specifically, a weak interaction decreases CO_2 selectivity, while a strong interaction increases the amount of energy required for regenerating material. In the rest of this chapter, physical and chemical properties of amines relevant to CO_2-capture technologies are briefly reviewed with a focus on molecular structures and interactions.

2.2   Classification of Amines

Amines fall into different classes depending on the number of the hydrogen atoms attached to the nitrogen atom. As illustrated in Fig. 2.1, primary and secondary amines contain nitrogen atoms covalently attaching two hydrogen atoms ($-NH_2$) and one hydrogen atom (>NH), respectively, while there are not any hydrogen atoms directly bound to the nitrogen in tertiary amines (>N–).

It is often mentioned that reactivity toward CO_2 decreases in the following order: primary ($-NH_2$) > secondary (>NH) > tertiary (>N–). However, this is not necessarily true. In order to properly compare different amines, you have to consider both electronic and steric effects of all substituents attached to the nitrogen atoms. You had better not give a reason for any experimental fact simply from the difference in class of amines.

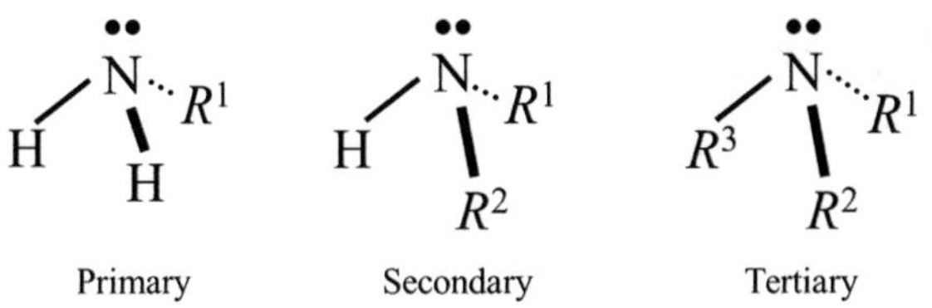

Fig. 2.1 Classification of amines

2.3 Boiling Point and Viscosity

As described in the above-mentioned patent, the boiling point of amine is one of the important physical properties for CO_2-capture applications. In particular, low-boiling-point amines are not favorable from environmental and economical aspects because they are likely to volatilize.

The boiling points of amines representative of CO_2-capture solvents are given in Table 2.1. In general, one can say that the higher molecular weight, the higher the boiling point. In addition to the molecular weight, intermolecular interactions such as electrostatic interaction greatly affect the boiling point. Because the hydroxy group and amino group can form relatively strong hydrogen bonds, alkanolamines have much higher boiling points than alkylamines with similar molecular weights.

The viscosity of amine is another important physical property for CO_2-capture performance. The diffusion coefficient of particles in the solvent is simply given by the well-known Stokes–Einstein relation:

$$D \propto T\mu^{-1} \tag{2.1}$$

where μ is the solvent viscosity and T is the absolute temperature. This relation is strictly valid in diluted colloidal particle systems only, but is remarkably applicable to a wide range of systems. According to relation (2.1), we can predict that the higher amine viscosity, the slower diffusion of CO_2 in the amine solvent. It means that low-viscosity amines are advantageous in terms of fast kinetics.

Table 2.1 Amines boiling point and viscosity[a]

Amine[b]	Molecular weight (g/mol)	Boiling point (°C)[c]	Viscosity (cP)
MEA	61.08	171	24[d]
DEA	105.14	268	380[e]
MDEA	119.16	247	101[d]

[a]Dow chemical data sheet
[b]Monoethanolamine; diethanolamine; N-methyldiethanolamine
[c]At 760 mm Hg
[d]At 20 °C
[e]At 30 °C

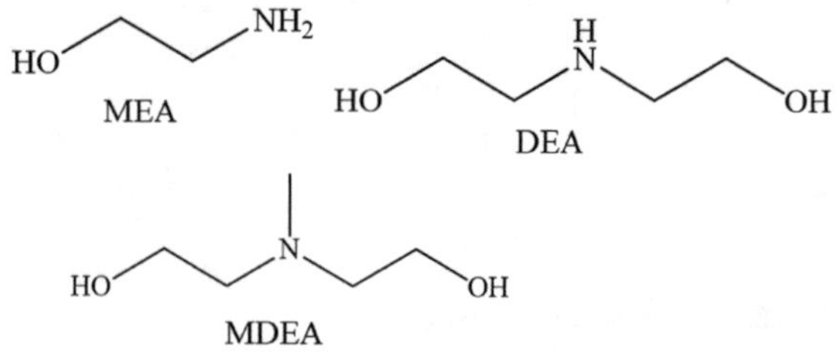

Fig. 2.2 Monoethanolamine, diethanolamine, and N-methyldiethanolamine

As is the case with the boiling point, attractive intermolecular interactions among amine molecules increase the solvent viscosity. Therefore, alkanolamines have much higher viscosities than alkylamines with similar molecular weights. As listed in Table 2.1, the viscosity of diethanolamine (DEA) is significantly higher than that of MEA, which can be explained by the member of the hydroxy groups (–OH) as shown in Fig. 2.2.

Of course, both the boiling point and viscosity represent physical properties of pure amines. These physical properties of pure amines are useful guides to understanding the structure–activity relationship. However, in actual CO_2 capture, amines are used in aqueous solutions, in porous materials, in polymeric membranes, or in other surroundings. This is partly because of the viscosity, volatility, and corrosivity of pure amines. Furthermore, anions and cations produced by chemical reaction between amines and CO_2 give complex and strong intermolecular interactions. Therefore, considerations for such surrounding environments may be required in some cases [23].

2.4 Activity and Concentration

Amine solvents for CO_2 capture usually contain a significant amount of amine. For example, the benchmark solvent consists of 30 wt% MEA and 70 wt% water. Furthermore, a considerable amount of charged species forms in the solvents by absorbing CO_2. Because such solutions may be far from an ideal solution, activity, i.e., "effective concentration" must be introduced for rigorous modeling. In this study, however, we assume all the activity coefficients to be 1 for simplicity. In other words, we use the molar concentration instead of activity. We also assume the molar concentration of water is constant.

2.5 Basicity of Amines

The central of CO_2–amine chemistry is acid–base reaction. Therefore, the basicity of amine is the most critical factor governing the CO_2-capture performance.

2.5.1 Definition of Base

A "base" is defined in several ways: An Arrhenius base is a species that increases the concentration of hydroxide anions (OH^-) in aqueous solutions; a Brønsted base is a species that can accept a proton (H^+); and a Lewis base is a species that reacts with a Lewis acid by donating its electron pair to form a Lewis adduct. Amines can act as both Brønsted bases and Lewis bases.

2.5.2 Amine pK_a

In aqueous solutions, an amine reacts with a proton as a Brønsted base, and the species distribution in the equilibrium is related to its pK_a.

$$R^1 R^2 R^3 NH^+ + H_2O \rightleftarrows R^1 R^2 R^3 N + H_3O^+ \tag{2.2}$$

$$pK_a = -\log([R^1 R^2 R^3 N][H_3O^+]/[R^1 R^2 R^3 NH^+]) \tag{2.3}$$

where R^n represents a hydrogen atom or any substituent, and K_a is the acid dissociation constant of the pronated amine. The value of pK_a defined by Eq. (2.3) indicates the extent of dissociation of the conjugate acid, $R^1 R^2 R^3 NH^+$. However, it is often simply described as "amine pK_a." The more positive the value of pK_a, the higher the Brønsted basicity of amine.

Table 2.2 lists the pK_a values of various amines with their molecular structures. A part of amines (ca. 99% purity) were purchased from various chemical companies and used without further purification. For them, the pK_a values in aqueous solutions were determined by potentiometric titration at room temperature with sodium hydroxide and hydrochloric acid solutions. For the rest of amines, the pK_a values were obtained from the literature [20].

A comparison of the amine pK_a values offers valuable insight. The order of pK_a values of MEA, DEA, and triethanolamine (TEA) is as follows: 9.53 (MEA) > 8.96

Table 2.2 pK_a value[a]

Amine	pK_a	Amine	pK_a
Monoethanolamine, MEA	9.53[b]	2-(Methylamino)ethanol, MAE	9.93[b]
Diethanolamine, DEA	8.96[c]	2-(Ethylamino)ethanol, EAE	9.99[b]
Triethanolamine, TEA	7.72[b]	2-amino-2-methyl-1-propanol, AMP	9.72[c]
2-Aminopyridine, AP	6.70[b]	N-methyldiethanolamine, MDEA	8.51[b]

[a] At 24–27 °C
[b] Yamada et al. [20]
[c] Perrin [16]

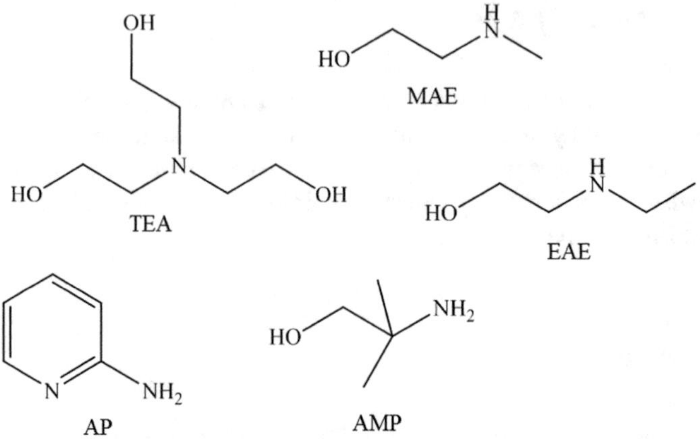

Fig. 2.3 Triethanolamine, 2-aminopyridine, 2-(methylamino)ethanol, 2-(ethylamino)ethanol, 2-amino-2-methyl-1-propanol, and *N*-methyldiethanolamine

(DEA) > 7.72 (TEA), indicating that the hydroxy group weakens the basicity of amino group through the sigma bonds (N–C–C–O) because of its electron-withdrawing nature. In contrast, the alkyl group enhances the basicity of amino group because of its electron-donating nature. For this reason, *N*-alkyl alkanolamines such as 2-(methylamino)ethanol (MAE) and 2-(ethylamino)ethanol (EAE) show higher pK_a values than those of ethanolamines with similar molecular weights. Aromatic amines show relatively low pK_a values, e.g., 6.70 for 2-aminopyridine (AP), of which the lone pair of electrons are delocalized in the simple aromatic ring (Fig. 2.3).

The amine pK_a is defined using the reaction between an amine and a proton. Therefore, it directly represents a Brønsted basicity. On the other hands, a Lewis basicity of amine is also important for analyzing the reactivity of amine toward CO_2, as discussed in detail below.

2.5.3 Comparison of Amine with Water

CO_2 also chemically reacts with pure water:

$$CO_2 + 2H_2O \rightleftarrows HCO_3^- + H_3O^+ \tag{2.4}$$

where one water molecule (H_2O) acts as a Brønsted base. However, the solubility of CO_2 in water is well correlated with a model based on Henry's law and is only 1.5 g/kg under the CO_2 partial pressure of 101.325 kPa at 25 °C. It decreases with an increase in temperature as shown in Fig. 2.4. Since amine solutions are more basic, such solutions absorb a significant amount of CO_2. For example, the CO_2 solubility exceeds 100 g/kg in aqueous solutions of 30 wt% MEA under similar conditions.

Fig. 2.4 CO_2 solubility in water calculated from a correlation model [4]

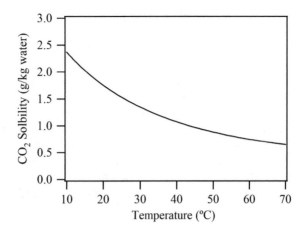

2.6   Carbamate and Bicarbonate Formations

Two major pathways that contribute to amine-based CO_2 capture in various solvents, adsorbents, or membranes include the formations of carbamate and bicarbonate anions. Primary and secondary amines react with CO_2 to form the carbamate anion and protonated amine, while primary, secondary, and tertiary amines react with CO_2 to form the bicarbonate anion and protonated amine:

$$R^1R^2NH + CO_2 + B \rightleftarrows R^1R^2NCOO^- + BH^+ \qquad (2.5)$$

$$R^1R^2R^3N + CO_2 + H_2O \rightleftarrows HCO_3^- + R^1R^2R^3NH^+ \qquad (2.6)$$

where B represents any Brønsted base in the system. Usually, the amine is abundant and the strongest Brønsted base in the system. Therefore, the following reaction is exclusively responsible for the carbamate formation:

$$2R^1R^2NH + CO_2 \rightleftarrows R^1R^2NCOO^- + R^1R^2NH_2^+ \qquad (2.7)$$

where two amine molecules take part in the capture of one CO_2 molecule. In this reaction, one amine molecule serves as a Brønsted base, while the other serves as a Lewis base.

2.6.1   Carbamate Stability

A stability of carbamate is evaluated by the amine carbamate stability constant K_C, which is the equilibrium constant for reaction (2.8).

$$HCO_3^- + R^1R^2NH \rightleftharpoons R^1R^2NCOO^- + H_2O \qquad (2.8)$$

$$K_C = [R^1R^2NCOO^-]/([HCO_3^-][R^1R^2NH]) \qquad (2.9)$$

Here, the more lager the value of K_C, the higher the Lewis basicity of amine. In other words, an amine with a high K_C value forms a stable carbamate.

Previously, we compared the carbamate stabilities between MEA and DEA based on quantum chemical calculations: $\log K_C = 1.31$ and 0.93 at 25 °C for MEA and DEA, respectively [20]. This difference in carbamate stability can be ascribed to the number of hydroxy groups. However, roles of the hydroxy group are not so straightforward. As described before, the hydroxy group weakens the basicity of amino group because of its electron-withdrawing nature. Besides, the hydroxy group and the adjacent ethylene group ($HO–CH_2CH_2–$) make the carbamate unstable due to the steric hindrance. Consequently, the carbamate of DEA has less stability than that of MEA. Simultaneously, the hydroxy group has the opposite effect: It stabilizes the carbamate anion by the intramolecular hydrogen bond $(-COO^- \cdots HO-)$ as shown in Fig. 2.5.

Compared to pK_a values, the amine carbamate stability constants are not easily available from experiments. Alternatively, quantum chemical approaches such as the density functional theory (DFT) are useful to predict the amine carbamate stability by calculating the Gibbs free energy [20]. The change in the free energy calculated for reaction (2.8) gives the carbamate stability from the relation between the reaction free energy and the equilibrium constant:

$$\Delta G_{(2.8)} = -RT \ln K_C \qquad (2.10)$$

where R is the gas constant. The calculated reaction free energy, $\Delta G_{(2.8)}$, for several amines is compared in Fig. 2.6. The calculations quantitatively show that the carbamate of 2-amino-2-methyl-1-propanol (AMP), which is known as a "sterically hindered amine," is significantly unstable compared with that of MEA or other mildly hindered secondary amines.

Fig. 2.5 Intramolecular hydrogen bond of DEA carbamate

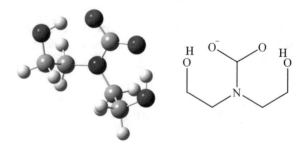

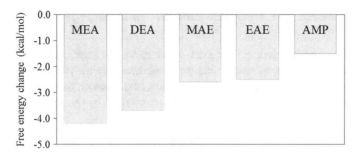

Fig. 2.6 Free energy change calculated for carbamate formation from bicarbonate via reaction (2.8) in aqueous amine solutions

2.6.2 Ratio of Carbamate to Bicarbonate

A branching ration between carbamate and bicarbonate formations reflects characteristics of the amine used for CO_2 capture. As shown in the above equations, amines, protonated amines, carbamate anions, bicarbonate anions, protons, and water exist in the CO_2-loaded aqueous solutions of primary and secondary amines. In addition to these, carbonate anions and hydroxide anions also coexist.

$$CO_3^{2-} + H_2O \rightleftharpoons HCO_3^- + OH^- \tag{2.11}$$

The ratio of carbonate to bicarbonate, $[HCO_3^-]/[CO_3^{2-}]$, decreases with a decrease in pH:

$$pH = -\log[H_3O^+] \tag{2.12}$$

where $[H_3O^+]$ is the concentration of proton in units of mol/L.

Spectroscopic methods including nuclear magnetic resonance (NMR) are useful for speciation and quantification in amine–CO_2–water systems. For instance, a ^{13}C NMR study showed that the yields of carbamates, defined by Eq. (2.13), were 4.3, 1.5, and 0.5 for MEA, DEA, and EAE, respectively, in their CO_2-loaded aqueous solutions (ca. 0.5–0.6 mol CO_2/mol amine) at 22 °C [20].

$$\Psi = [R^1R^2NCOO^-]/([R^1R^2NCOO^-] + [HCO_3^-] + [CO_3^{2-}]) \tag{2.13}$$

A conventional explanation for the above results is that the carbamates of secondary amines are less stable than the carbamate of primary amine MEA, because of the steric hindrances of secondary amines. However, the difference between DEA and EAE cannot be explained only by the steric effect. As noted before, you have to consider electronic effects as well as steric effects of the substituents attached to the nitrogen atom.

2.6.3 Kinetics of Carbamate and Bicarbonate Formations

Generally, carbamate formation shows faster retraction kinetics than bicarbonate formation. A lot of experimental kinetic data are available for various amines under a wide range of conditions in the literature. Here, we focus on the activation energy, E_a, which is directly related to the reaction rate constant, k, by the Arrhenius equation:

$$k = A \exp\{-E_a/(RT)\} \tag{2.14}$$

where A is the pre-exponential factor for the reaction. The activation energy in this empiric formula can be understood as the minimum energy required for conversion from the reactant system to the product system as schematically depicted in Fig. 2.7. Eyring's transient-state theory (TST) rigorously formulates the relation between the activation energy and the pre-exponential factor.

The quantum chemical methods are also useful to calculate the activation energy by accessing transient states of elementary reactions, in which the molecular geometries are determined within the framework of TST. Figure 2.8 shows the transition state geometry of rate-determining step in MEA carbamate formation, optimized at the DFT(B3LYP)/6-311++G(d,p) level of theory. The implicit continuum solvation model (SMD/IEF-PCM) is used to reflect the effect of surrounding solvent with the dielectric constant of water [23]. The product of this rate-determining step is a zwitterion.

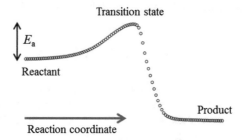

Fig. 2.7 Schematic of activation energy, E_a

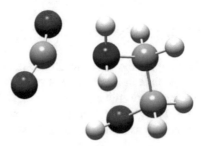

Fig. 2.8 Transition state of reaction of CO_2 with MEA

Table 2.3 Activation energies (kcal/mol)

Amine	Carbamate formation	Bicarbonate formation
MEA[a]	6.6	15.3
AMP[b]	7.6	15.8

[a]PCM/CCSD(T)/6-311++G(2df,2p)//PCM/MP2/6-311++G(d,p) [21]
[b]CPCM/CCSD(T)/6-311++G(d,p)//SMD/IEF-PCM/B3LYP/6-31G(d) [14]

$$HO(CH_2)_2NH_2 + CO_2 \rightleftarrows HO(CH_2)_2NH_2^+COO^- \tag{2.15}$$

The calculated activation energies of carbamate and bicarbonate formations in aqueous amine (MEA or AMP) solutions are compared in Table 2.3. For these comparisons, the rate-determining step, i.e., zwitterion formation, is analyzed to calculate the activation energy of carbamate, while a single-step termolecular mechanism for reaction (2.6) is analyzed to calculate the activation energy of bicarbonate formation.

As experimentally shown and also theoretically predicted (Fig. 2.6), the main products of the absorption reaction of CO_2 in aqueous amine solutions are carbamate and bicarbonate for MEA and AMP, respectively. However, it is theoretically predicted that for both the amines, the carbamate anion forms via a zwitterion intermediate, and its activation energy is lower than that of the bicarbonate formation (Table 2.3). Quantum chemical methods provide such a valuable insight that is difficult to obtain only from experiments.

2.7 Other Species Formed by Amine–CO_2 Reaction

2.7.1 Carbamic Acid

Another possible reaction product of amine–CO_2 reaction is the carbamic acid.

$$R^1R^2NH + CO_2 \rightleftarrows R^1R^2NCOOH \tag{2.16}$$

The carbamic acid would be a minor product if produced because usual materials for CO_2 capture such as aqueous amine solutions are under basic conditions that would easily accept the proton from the carbamic acid to yield the carbamate anion. Furthermore, as discussed later, most materials for CO_2 capture stabilize ionic products such as carbamate and protonated amine by their polarity. Therefore, the charged products, i.e., carbamate and proton, are more stable in most instances.

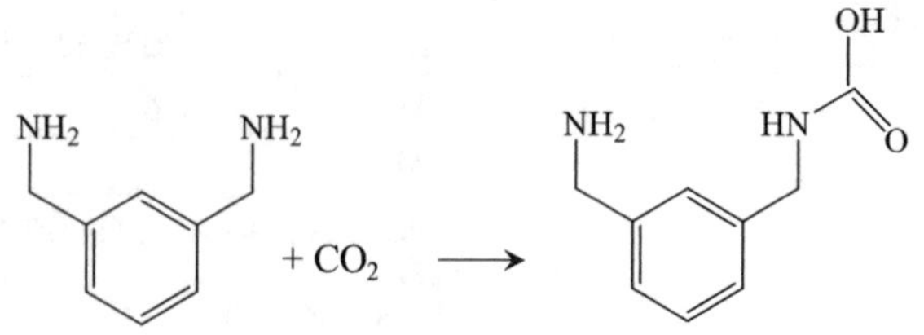

Fig. 2.9 Carbamic acid formation from *m*-xylylenediamine and CO_2 [13]

However, by tuning the polarities of amine or its surroundings, it is possible to significantly produce the carbamic acid. For instance, carbamic acid formations have been reported for amine-grafted mesoporous silica materials [2]. Also in liquid phase, some hydrophobic amines such as *m*-xylylenediamine react with CO_2 to form the carbamic acids as shown in Fig. 2.9 [13].

2.7.2 Alkyl Carbonate

Alkanolamines can produce alkyl carbonates by the reaction of their hydroxy groups with CO_2.

$$R^1 R^2 N R^3 OH + CO_2 + B \rightleftarrows R^1 R^2 N R^3 OCOO^- + BH^+ \qquad (2.17)$$

The alkyl carbonate is likely to form in the absence of water instead of bicarbonate. Therefore, it is often observed in non-aqueous alkanolamine solutions. However, it has been identified even in aqueous amine solutions and under ambient temperature and pressure condition, although it is a minor product. We have confirmed that a small amount of CO_2 is covalently bonded to hydroxyl groups of EAE carbamate as well as EAE in aqueous solutions by 2D NMR spectroscopy DFT calculations [22].

2.8   Mechanism of Amine–CO_2 Reaction

An overall chemical reaction contains one or more elementary reactions. An elementary reaction proceeds in a single-step reaction with a single transition state. In this section, elementary reactions in the overall reactions (2.6), (2.7), and (2.8) are discussed.

For instance, it had been controversial whether a reaction intermediate existed or not in the carbamate formation reaction (2.7). If an intermediate exists with a certain lifetime that is appreciably longer than a molecular vibration, it is not a single-step reaction (Fig. 2.10). Such an intermediate is difficult to identify from experiments because of its short lifetime. As a complementary approach, quantum mechanical studies have shed light on the reaction mechanism of amine–CO_2.

Fig. 2.10 Schematic energy diagram for two-step reaction

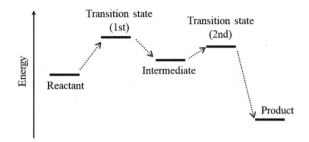

The carbamate formations in aqueous MEA solution have been well explored by such theoretical methods. Some of these studies have concluded that the MEA zwitterion produced by reaction (2.15) is not merely a transient species, but can exist as an intermediate. Therefore, the MEA carbamate formation is not a single-step reaction.

In reality, however, reaction mechanisms are determined by various factors including the composition of solvent and solvation effect. Also, reactions may proceed in more than one mechanisms in parallel. In this chapter, therefore, we avoid the "either/or" thinking and look at several possible mechanisms.

2.8.1 Zwitterion Mechanism

In the zwitterion mechanism, the carbamate formation reaction (2.7) proceeds in two steps. In the first step, a zwitterion forms:

$$R^1R^2NH + CO_2 \rightleftarrows R^1R^2NH^+COO^- \tag{2.18}$$

and then, in the next step the zwitterion is deprotonated by any Brønsted base.

$$R^1R^2NH^+COO^- + B \rightleftarrows R^1R^2NCOO^- + BH^+ \tag{2.19}$$

Some DFT studies combined with solvation models have supported this mechanism in aqueous solutions of amine such as MEA [14, 19] and AMP [21]. These studies also have confirmed that the formation of the zwitterion (2.18) is rate determining: In other words, the following proton transfer (2.19) occurs through an energy barrier lower than that for the first-step reaction.

2.8.2 Carbamic Acid Mechanism

The first step in the carbamate formation reaction (2.7) is considered to take place between a nonbonding electron pair at the amino nitrogen atom and an antibonding empty orbital in CO_2 for a donor–acceptor. One possible step is a zwitterion form, while another one is carbamic acid formation:

$$R^1R^2NH + CO_2 \rightleftarrows R^1R^2NCOOH \qquad (2.20)$$

which is bimolecular, of second order, and rate determining, while the second proton transfer step to Brønsted bases such as amine and water is assumed to be instantaneous.

DFT and high-level ab initio calculations for aqueous MEA and DEA solutions have partly supported the carbamic acid mechanism [1]. It should be noted that this study has been performed by energy calculations based on the molecular geometry optimized in vacuum. As mentioned above, tuning the polarities of amine or its surroundings, it is possible to significantly produce the carbamic acid. Due to the same reason, a carbamic acid intermediate can be more stable than a zwitterion intermediate in less polar media.

2.8.3 Single-Step Termolecular Mechanism

Crooks and Donnellan [5] proposed the single-step carbamate formation mechanism as illustrated in Fig. 2.11, based on their kinetic experiments for the reaction of CO_2 in aqueous solutions of various amines and concluded that the much-quoted Danckwerts mechanism [8], i.e., zwitterion intermediate, was shown to be unlikely.

Entering the twenty-first century, da Silva and Svendsen [6] introduced quantum mechanical calculations into the amine–CO_2 chemistry. They studied the formation process of carbamate from CO_2 and MEA and suggested that the zwitterion may be an entirely transient state because no stable zwitterion species was formed.

A study using DFT and coupled-cluster theory calculations has indicated that the zwitterion of highly basic amine may have a longer lifetime, in contrast to that of MEA, where it is only a transient species [15].

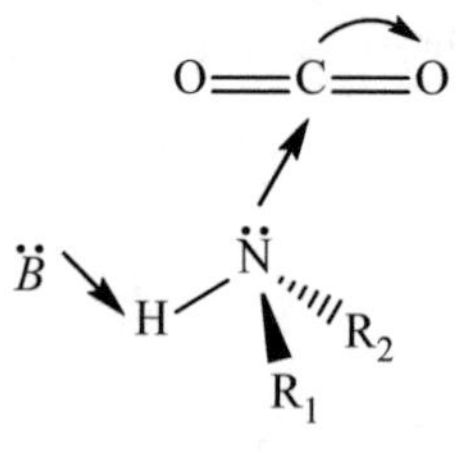

Fig. 2.11 Single-step carbamate formation mechanism

2.8.4   Base-Catalyzed Hydration

Donaldson and Nguyen [9] studied reaction kinetics of CO_2 with TEA and triethylamine in aqueous membranes. They suggested that the reaction mechanism for TEA was the base catalysis of CO_2 hydration (Fig. 2.12), while triethylamine appeared to act only as a weak base to produce free OH^- that reacts with CO_2.

Quantum mechanical calculations for the above base-catalyzed mechanism were performed by da Silva and Svendsen [7] with one H_2O molecule and one CO_2 molecule in the presence of an MEA molecule and the transition-state geometry was determined, revealing the single-step termolecular mechanism by H_2O, CO_2, and MEA molecules.

Our DFT studies also have revealed the single-step termolecular mechanism for bicarbonate formations in aqueous solutions of MEA [14] and AMP [21]. Figure 2.13 shows the transition state of single-step bicarbonate formation by H_2O, CO_2, and AMP molecules, which is optimized at the DFT(B3LYP)/6-31G(d) level of theory combined with the solvation model (SMD/IEF-PCM) for water.

The direct reaction of CO_2 with the hydroxide anion is also possible to occur to form the bicarbonate.

$$CO_2 + OH^- \rightleftharpoons HCO_3^- \tag{2.21}$$

Usually, the energy barrier for reaction (2.21) is much lower than that for the termolecular mechanism, whereas the concentration of hydroxide anion is much lower

Fig. 2.12 Base catalysis of CO_2 hydration

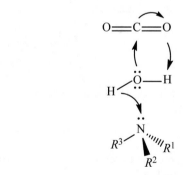

Fig. 2.13 Transition state of reaction of CO_2, H_2O and AMP

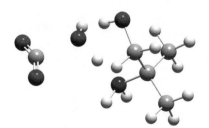

than that of amine. Thus, reaction (2.21) substantially contributes to the bicarbonate formation under high pH conditions.

2.8.5   Interconversion of Carbamate and Bicarbonate

It has been experimentally proven that aqueous solution of AMP, which is a sterically hindered primary amine, absorbs CO_2 to predominantly produce the bicarbonate anion at equilibrium. A conventional explanation for the route to this product is the hydrolysis of the AMP carbamate because of its instability.

$$HOCH_2C(CH_3)_2NHCOO^- + H_2O \rightleftarrows HCO_3^- + HOCH_2C(CH_3)_2NH_2 \qquad (2.22)$$

This equation represents an overall reaction. However, it has often been mistaken as an elementary reaction. We performed intrinsic reaction coordinate (IRC) analyses in the aqueous solution phase by DFT for this system using the continuum solvation model (SMD/IEF-PCM) and compared the activation energies calculated for various elementary reactions [21]. As a result, we found that reaction (2.22) hardly occurred as an elementary reaction because its high activation energy was too high.

However, the interconversion of carbamate and bicarbonate may take place in the presence of catalyst. We also carried out ab initio molecular orbital calculations combined with the continuum solvation model (PCM) for the CO_2 absorption in aqueous MEA solutions [14]. The calculations identified an alternative pathway for hydrolysis. First, a proton transfer from protonated MEA to carbamate to generate the carbamic acid:

$$HO(CH_2)_2NHCOO^- + HO(CH_2)_2NH_3^+$$
$$\rightleftarrows HO(CH_2)_2NHCOOH + HO(CH_2)_2NH_2 \qquad (2.23)$$

which is followed by nucleophilic addition of OH^- to the carbamic acid:

$$HO(CH_2)_2NHCOOH + OH^- \rightleftarrows HO(CH_2)_2N(OH)_2O^- \qquad (2.24)$$

and then, the subsequent low-barrier reaction leads to the formation of bicarbonate and MEA.

$$HO(CH_2)_2N(OH)_2O^- \rightleftarrows HCO_3^- + HO(CH_2)_2NH_2 \qquad (2.25)$$

2.9 Effect of Reaction Field

In this chapter, the reaction of amine with CO_2 is reviewed mainly focusing on the aqueous phase conditions. As described before, the effects of surrounding environments are also of importance. The major products and the reaction intermediates contain charged species such as carbamate anion, bicarbonate anion, and protonated amine.

The Born formula gives the solvation energy of a charged species depending on the dielectric constant ε_r:

$$E_{solv} = \left[N_A Z^2 e^2 / (8\pi \varepsilon_0 a) \right] \left(1 - \varepsilon_r^{-1} \right) \tag{2.26}$$

where N_A is the Avogadro number, ε_0 is the permittivity of free space, and a is a spherical radius of the species with charge Ze. The dielectric constant is a measure of polarity, and it greatly affects the energy of charged species according to this formula. Therefore, the energy diagram of reaction involving charged species is potentially governed to a large extent by the polarity of the reaction field.

Table 2.4 summaries the dielectric constants of various solvents. Needless to say, water, that is a solvent with high polarity, greatly stabilizes the above-mentioned charged species relevant to the reaction of amine with CO_2. Note that water molecules also stabilize the charged species by the hydrogen-bond effect. Some solvation models also treat such hydrogen-bond effects implicitly or explicitly.

We analyzed reaction (2.7) for MEA by DFT combined with the continuum solvation model (SMD/IEF-PCM) by changing the dielectric constant. As is well known, the carbamate formation of MEA is exothermic. However, the calculation results predicted that it shifts to endothermic with a decrease in the dielectric constant. The activation energy for the rate-determining step, i.e., the zwitterion formation

Table 2.4 Dielectric constants of various solvents[a]

Solvent	ε_r	Solvent	ε_r
Formamide	108.94	1-Octanol	9.86
Water	78.355	Propyl amine	4.99
Ethylene glycol	40.245	Butyl amine	4.62
N,N-Dimethylacetamide	37.781	Diethyl ether	4.24
N,N-Dimethylformamide	37.219	Pentyl amine	4.20
Acetonitrile	35.688	Diphenyl ether	3.73
Methanol	32.613	Diethyl amine	3.58
Ethanol	24.852	Dipropyl amine	2.91
Acetone	20.493	Toluene	2.37
Benzyl alcohol	12.457	1,4-Dioxane	2.21

[a]Frisch et al. [11]

also drastically changed depending on the dielectric constant, which indicated that the MEA carbamate could not generate spontaneously in low-polar media [23]. As mentioned before, carbamic acid is likely to form in such low-polar reaction fields including gas phase.

As described before, the higher solvent viscosity, the lower diffusivity of CO_2 gas. Therefore, the CO_2 absorption rate is limited in high-viscosity solvents. Also, aqueous amine solutions typically increase their viscosities by absorbing CO_2 because the production of charged ions leads to strong intermolecular interactions. Therefore, a deceleration in CO_2 absorption of aqueous amine solutions with increasing CO_2 loading is explained by the increased solvent viscosity as well as the decreased concentration of flesh amines. Furthermore, the above disunions on the decreased dielectric constant suggest that the decreased dielectric constant of solutions with CO_2 loading contributes to the deceleration in CO_2 absorption.

Interestingly, the drastic dependence of reactivity on the dielectric constant observed for aqueous MEA solutions is not the case with the following type of reaction:

$$R^1 R^2 R^3 N^- + CO_2 \rightleftarrows R^1 R^2 R^3 NCOO^- \qquad (2.27)$$

where $R^1 R^2 R^3 N^-$ represents the amine-functionalized anion in ionic liquids, e.g., the glycine anion [23]. The calculations have shown that in contrast to aqueous amine solutions, the energy diagram of the reaction between CO_2 and amine-functionalized anions barely depends on the dielectric constant of surrounding continuum. The formed CO_2-bound anion is likely to be stabilized in ILs regardless of the dielectric constant.

2.10 Conclusion

Physical and chemical properties of amines and amine-containing media for CO_2 capture have been reviewed in this chapter. The physical properties including the boiling point, diffusion coefficient, and polarity are of vital information for understanding, designing, and optimizing amine-based CO_2-capture technologies. In those technologies, the central chemical reactions are carbamate formations and bicarbonate formations. The branding ratio of these reactions, which depends on the structure of amine molecule, greatly affects the CO_2-capture performance. The complicated molecular interactions, various elementary steps, by-products, and surrounding envelopments are involved in these reactions. Among them, the effects of substituents such as alkyl and hydroxy groups, hydrogen bond, and solvation are of paramount importance.

References

1. Arstad B, Blom R, Swang O (2007) CO_2 absorption in aqueous solutions of alkanolamines: mechanistic insight from quantum chemical calculations. J Phys Chem A 111:1222–1228
2. Bacsik Z, Ahlsten N, Ziadi A, Zhao G, Garcia-Bennett AE, Martín-Matute B, Hedin N (2011) Mechanisms and kinetics for sorption of CO_2 on bicontinuous mesoporous silica modified with n-propylamine. Langmuir 27:11118–11128
3. Bottoms RR (1930) Separating acid gases. U.S. Patent 1783901
4. Carroll JJ, Slupsky JD, Mather AE (1991) The solubility of carbon dioxide in water at low pressure. J Phys Chem Ref Data 20:1201–1209
5. Crooks JE, Donnellan JP (1989) Kinetics and mechanism of the reaction between carbon dioxide and amines in aqueous solution. J Chem Soc Perkins Trans II, 331–333
6. da Silva EF, Svendsen HF (2004) Ab initio study of the reaction of carbamate formation from CO_2 and alkanolamines. Ind Eng Chem Res 43:3413–3418
7. da Silva EF, Svendsen HF (2007) Computational chemistry study of reactions, equilibrium and kinetics of chemical CO_2 absorption. Int J Greenhouse Gas Control 2:151–157
8. Danckwerts PV (1979) The reaction of CO_2 with ethanolamines. Chem Eng Sci 34:443–446
9. Donaldsen TL, Nguyen YN (1980) Carbon dioxide reaction kinetics and transport in aqueous amine membranes. Ind Eng Chem Fundam 19:260–266
10. Dutcher B, Fan M, Russell AG (2015) Amine-based CO_2 capture technology development from the beginning of 2013—a review. ACS Appl Mater Interfaces 7:2137–2148
11. Frisch MJ, Trucks GW, Schlegel HB, Scuseria GE, Robb MA, Cheeseman JR, Scalmani G, Barone V, Mennucci B, Petersson GA, Nakatsuji H, Caricato M, Li X, Hratchian HP, Izmaylov AF, Bloino J, Zheng G, Sonnenberg JL, Hada M, Ehara M, Toyota K, Fukuda R, Hasegawa J, Ishida M, Nakajima T, Honda Y, Kitao O, Nakai H, Vreven T, Montgomery JA Jr, Peralta JE, Ogliaro F, Bearpark M, Heyd JJ, Brothers E, Kudin KN, Staroverov VN, Kobayashi R, Normand J, Raghavachari K, Rendell A, Burant JC, Iyengar SS, Tomasi J, Cossi M, Rega N, Millam JM, Klene M, Knox JE, Cross JB, Bakken V, Adamo C, Jaramillo J, Gomperts R, Stratmann RE, Yazyev O, Austin AJ, Cammi R, Pomelli C, Ochterski JW, Martin RL, Morokuma K, Zakrzewski VG, Voth GA, Salvador P, Dannenberg JJ, Dapprich S, Daniels AD, Farkas Ö, Foresman, JB, Ortiz JV, Cioslowski J, Fox DJ (2009) Gaussian 09, Revision E.01, Gaussian, Inc. Wallingford CT
12. Hayes KS (2001) Industrial processes for manufacturing amines. Appl Catal A 221:187–195
13. Inagaki F, Matsumoto C, Iwata T, Mukai C (2017) CO_2-selective absorbents in air: reverse lipid bilayer structure forming neutral zcarbamic acid in water without hydration. J Am Chem Soc 139:4639–4642
14. Matsuzaki Y, Yamada H, Chowdhury FA, Higashii T, Onoda M (2013) Ab Initio study of CO_2 capture mechanisms in aqueous monoethanolamine: reaction pathways for the direct Interconversion of carbamate and bicarbonate. J Phys Chem A 117:9274–9281
15. Orestes E, Ronconi CM, Carneiro JWM (2014) Insights into the interactions of CO_2 with amines: A DFT benchmark study. Phys Chem Chem Phys 16:17213–17219
16. Perrin DD (1965) Dissociation constants of organic bases in aqueous solution. Butterworths, London. (1972) Supplement
17. Piro A, Tagarelli G, Lagonia P, Tagarelli A, Quattrone A (2010) Casimir Funk: his discovery of the vitamins and their deficiency disorders. Ann Nutr Metab 57:85–88
18. Rochelle GT (2009) Amine scrubbing for CO_2 capture. Science 325:1652–1654
19. Xie H-B, Zhou Y, Zhang Y, Johnson JK (2010) Reaction mechanism of monoethanolamine with CO_2 in aqueous solution from molecular modeling. J Phys Chem A 14:11844–11852
20. Yamada H, Shimizu S, Okabe H, Matsuzaki Y, Chowdhury FA, Fujioka Y (2010) Prediction of the basicity of aqueous amine solutions and the species distribution in the amine–H_2O–CO_2 system using the COSMO-RS method. Ind Eng Chem Res 49:2449–2455
21. Yamada H, Matsuzak Y, Higashii T, Kazama S (2011) Density functional theory study on carbon dioxide absorption into aqueous solutions of 2-amino-2-methyl-1-propanol using a continuum solvation model. J Phys Chem A 115:3079–3086

22. Yamada H, Matsuzaki Y, Goto K (2014) Quantitative spectroscopic study of equilibrium in CO_2-loaded aqueous 2-(ethylamino)ethanol solutions. Ind Eng Chem Res 53:1617–1623
23. Yamada H (2016) Comparison of solvation effects on CO_2 capture with aqueous amine solutions and amine-functionalized ionic liquids. J Phys Chem B 120:10563–10568

Chapter 3
CO$_2$ Capture with Absorbents

Abstract This chapter presents CO$_2$ capture with absorbents, especially a gas separation using a solvent, "absorption." Gas separation of absorption is a commercialized technology in industries such as natural gas production and fertilizer plants. Also, the new application field emerges in these days. That is CO$_2$ capture for CCS (CO$_2$ capture and storage), which is a promising solution to the greenhouse gas issue. The chapter consists of seven subchapters: General theoretical description is located at the beginning (Sect. 3.1). Process flows for post-combustion CO$_2$ capture and pre-combustion CO$_2$ capture are shown in Sect. 3.2. CO$_2$ capture for CCS and its current research and development (R&D) will be traced in Sects. 3.3 and 3.4. Research Institute of Innovative Technology for the Earth (RITE) leads R&D of CO$_2$ capture with absorbents and develops high-performance solvents for CO$_2$ capture from a blast furnace gas in steelworks. Section 3.5 will outline the RITE's activities. RITE also contributes to the practical application of CO$_2$-capture technology to various emission sources in industries and spawn a new R&D field (Sect. 3.6). The last subchapter of Sect. 3.7 summarizes topics of CO$_2$ capture with absorbents.

Keywords Absorption · Amine · Post-combustion CO$_2$ capture · Pre-combustion CO$_2$ capture · Industrial CCS

3.1 Absorption

Absorption is a unit process in which a specific gas is separated from gas mixture using a solvent. It is useful for decreasing impurity to increase product concentration, removing toxic gaseous material from a gas stream, and so on. Well-known examples are such industrial processes as acid gas removal by alkaline aqueous solution, recovery alcohol vapor by water, and separation of hydrocarbons by hydrocarbon oil. Absorption is generally classified into two techniques based on dissolution mechanisms of gas molecule: physical absorption and chemical absorption.

© The Author(s), under exclusive license to Springer Nature Switzerland AG 2019   23
S. Nakao et al., *Advanced C O$_2$ Capture Technologies*,
SpringerBriefs in Energy, https://doi.org/10.1007/978-3-030-18858-0_3

3.1.1   Physical Absorption

Physical absorption is a method of physically dissolving a gas in a physical solvent. It is generally considered to be based on Henry's law, which states that a dissolved gas loading in a physical solvent is proportional to the partial pressure of the gas. As shown in the gas–liquid equilibrium diagram (Fig. 3.1a), the physical solvent absorbs a given gas under conditions of high partial pressure in gas phase and desorbs it under the condition of lower partial pressure. Since a given gas is separated from the solvent using a driving force of a pressure difference, it is possible to remove and recover the gas with less energy.

3.1.2   Chemical Absorption

Chemical absorption is a method in which a target gas reacts with an absorbent in a chemical solvent and is dissolved as a reaction product. As shown in Fig. 3.1b, the gas–liquid equilibrium exhibits high absorption loading even under the condition of low gas partial pressure at low temperature. The chemical solvent containing a solute gas can recover the solute gas and regenerate the absorption liquid by shifting a reaction equilibrium by heating operation. Generally, even when the target gas has a low concentration, it has high selectivity, but desorption by heating requires a large amount of thermal energy.

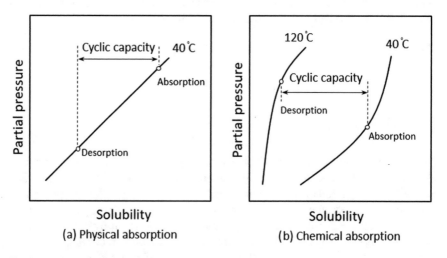

Fig. 3.1 Relation between solubility and partial pressure

3.2   Process Flows of Chemical Absorption

Process flows of chemical absorption will be illustrated with an example of CO_2 capture. An amine compound is a suitable absorbent for CO_2 capture and an amine-based solvent is commercially used in industries. CO_2 capture using an amine-based solvent has various process flows depend on process specification including pressure and concentration of a gas mixture. The following describes two cases: One is a process flow for CO_2 capture from atmospheric pressure gas mixture, and the other is a process flow for CO_2 capture from a pressurized gas mixture. They are generally called as "post-combustion CO_2 capture" and "pre-combustion CO_2 capture," respectively.

3.2.1   Atmospheric Pressure Gas (Post-combustion CO_2 Capture)

Figure 3.2 illustrates a typical chemical absorption process of CO_2 capture from atmospheric pressure gas mixture. In the process, a gas mixture and a solvent are contacted with countercurrent flow at the absorber. CO_2 absorbed solvent, "rich solvent," flows from the absorber bottom to the regenerator (stripper) top by way of a rich/lean heat exchanger. The temperature of the rich solvent at the regenerator is 100–120 °C. In the regenerator, the rich solvent is heated by steam generated at the reboiler. CO_2 is desorbed from the solvent, and then, the "lean solvent" is recirculated to the absorber. Chemical absorption for the atmospheric pressure gas mixture requires a large amount of thermal energy for solvent regeneration.

3.2.2   Pressurized Gas (Pre-combustion CO_2 Capture)

Chemical absorption is utilized for gas separation from a pressurized gas mixture, too. For example, natural gas production, CO_2 removal from syngas after water-gas-shift reaction, and so on. In the case of high pressurized gas mixture, the conventional process shown in Fig. 3.2 can be applied, but considering the gas pressure difference between absorption and desorption processes, the dissolved gas component dissolved in a solvent is easily desorbed to the gas phase by decreasing a pressure. The simplest process flow is to install a flash drum instead of a regenerator (Fig. 3.3). In practical use in the chemical absorption in the high-pressure system, optimization of the process related to target product and energy management is conducted. A system combining a flash drum and a regenerator with a reboiler and devising a method of circulating the solvent to meet a process target is another typical process flow (Fig. 3.4).

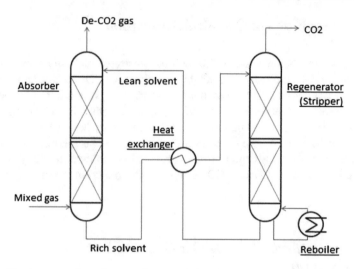

Fig. 3.2 Simple process flow diagram of chemical absorption (thermal regeneration)

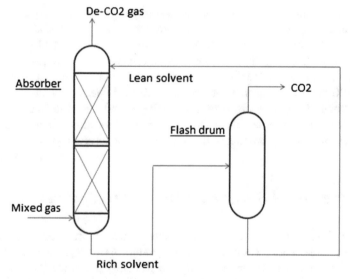

Fig. 3.3 Process flow diagram of chemical absorption with flash drum (regeneration by pressure decrease)

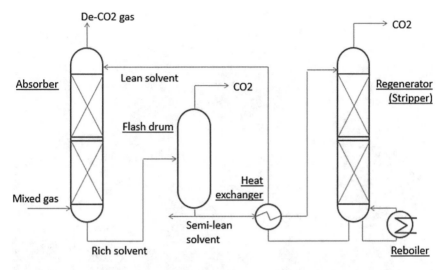

Fig. 3.4 Process flow diagram of chemical absorption with flash drum and regenerator

3.3 CO₂ Capture for CCS

CO_2 capture and storage (CCS), as well as clean energy technologies such as power generation from solar, wind and other renewable energies, is a promising solution to the greenhouse gas issue. In the system of CCS, CO_2 is captured from a large CO_2 emission source, compressed, transported, and finally injected into an underground storage area such as a deep saline layer. As shown in Fig. 3.5, the potential benefit of CCS on mitigating global greenhouse gas emissions is expected to reach 14% in 2050, according IEA Energy Technology Perspective [8]. This CCS scenario considered not only the power sector such as coal-fired and gas-fired power plants, but also energy intensive industries. Table 3.1 shows the large stationary CO_2 sources with emission of more than 0.1 MtCO₂ per year and clearly indicates that industrial sector can be accounted in the list [13].

3.3.1 Power Sector

Modern society relies heavily on fossil fuels. Especially, coal is one of the primary energy sources in the world, and it produces a large amount of CO_2, the major compound implicated in global warming. According to IEA' World Energy Outlook [9], coal represented 41% of the world's total electric power generation in 2016. In that year, CO_2 emissions from fuel combustion throughout the world reached 32.3 Gt, and contribution of coal combustion on it was 14.3 Gt [10]. Accordingly, coal combustion for electric power generation and associated greenhouse gas emissions

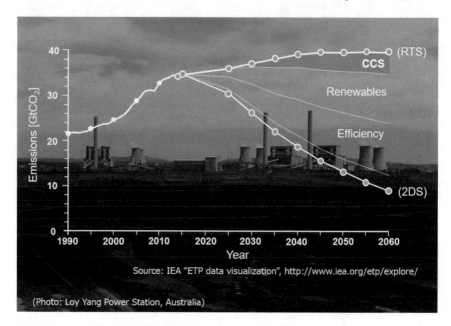

Fig. 3.5   CO$_2$ emission mitigation scenario

Table 3.1   Profile by process or industrial activity of worldwide large stationary CO$_2$ sources with emission of more than 0.1 MtCO$_2$ per year [13]

Process	Number of sources	Emissions (MtCO$_2$ yr^{-1})
Fossil fuels		
Power	4942	10,539
Cement production	1175	932
Refineries	638	798
Iron and steel industry	269	646
Petrochemical industry	470	379
Oil and gas processing	N/A	50
Other sources	90	33
Biomass		
Bioethanol and bioenergy	303	91
Total	7887	13,466

have become significant topics for technology developments. Specifically, much attention is paid on CO$_2$ capture from a flue gas of a coal-fired power plant in order to close to practical use of CCS.

3.3.2 Industry Sector

There is a large possibility to implement CO$_2$ capture for CCS in the industrial sector. The reason is that CO$_2$ capture is established as a commercial process in industries such as natural gas processing, ammonia, and urea production. CO$_2$ captured in these processes is released to atmosphere or provided to chemical and food industries. As shown in Table 3.1, the IPCC listed up cement production, refinery, iron and steel industry, petrochemical industry, and oil and gas processing. In the latest decade, as well as CO$_2$ capture from power plants, energy intensive industries have a strong interest in CO$_2$ mitigation technologies.

3.4 Research and Development

CO$_2$ capture for CCS is firstly reviewed in the IPCC report [13] and, after its publishing, CO$_2$ capture becames a hot topic in research and development worldwide. About ten years later from the report, we could see not a few new review papers on chemical absorption in journals and books. Some of them will be introduced below.

- Liang et al. [4] described broadly on the recent progress and new developments in post-combustion CO$_2$ capture with amine-based solvents. Their review includes solvent chemistry such as CO$_2$ solubility, reaction kinetics and NMR analysis, process design, solvent management issue of corrosion, and solvent stability.
- Mondal et al. [6] shows R&D activities of developing alternative amine-based solvents.
- Shi et al. [15] and Rayer et al. [14] presented information on solvent chemistry of amine–CO$_2$–H$_2$O. Experimental and theoretical studies of the CO$_2$ solubility in a chemical solvent were reviewed.
- Tan et al. [16] attempted to show key solvent technologies to develop an ideal chemical solvent from the viewpoints of process and plant specification.
- Idem et al. [7] showed the practical experiences in post-combustion CO$_2$ capture using chemical absorption, especially R&D activities using pilot and demonstration plants.
- New concepts of CO$_2$ capture were emerged in the technologies using solvents (IEAGHG 2014 [20]). Phase change solvent, ionic liquid, catalyzed solvent, and so on were classified into this movement.
- Review reports of CO$_2$ capture in industrial sectors can be found in the IEAGHG technical reports [11, 12].

3.5   State-of-the-Art Technology

Since the Cost-saving CO_2-Capture System (COCS) project (funded by the Ministry of Economy, Trade and Industry (METI), Japan), RITE conducted R&D on CO_2 capture from a blast furnace gas in the iron and steel industry. In COURSE 50 project (entrusted by the National Energy Research Institute for New Energy and Industrial Technology Development (NEDO), Japan) which began in 2008 [18], R&D in RITE gained special attention as developing CO_2-capture technology was responsible for two-thirds of the overall CO_2 mitigation target.

3.5.1   Development of New Solvents

RITE conducted R&D on CO_2 capture from a blast furnace gas in iron and steel indus-try and succeeded novel absorbents. High-performance solvents of CO_2 capture for CCS should have such features as low absorption heat, moderately high absorption rate, and large capacity of CO_2 capture. Figure 3.6 shows a schematic for the develop-ment of novel amine-based solvents. Initially, several hundred aqueous solutions of commercial amines were investigated to understand which characteristics lead to suf-ficient CO_2 capture. Then, the reaction rate with CO_2, CO_2-absorption capacity, and absorption heat were measured using a screening apparatus and a calorimetry instru-ment. Furthermore, synthetic absorbents were provided by clarifying the reaction mechanism between CO_2 and amines through analytical study and computational chemistry. Novel mixtures of amine compounds in which each amine compensates for the deficiencies of the others could be proposed based on the solvent data. That is, as well as selection of amine compounds, formulation of a chemical solvent with absorbents was a significant R&D approach to obtain the excellent performance of lower thermal energy requirement in comparison with a conventional solvent.

   Moreover, the technology development from the viewpoint of practical application and process optimization is very important. Because amines are exposed to impurities in combustion flue gas, degradation of amine absorbents, generation of by-products from amines, evaporation of chemicals, and their influence on the environment are important. That is, technology developments such as a reclaiming process to remove Heat Stable Salt (HSS) and an anti-corrosion agent are critical.

   In addition, suitable plant configurations vary with the amine-based solvent uti-lized. It is necessary to design an optimal process and the process simulation tech-nology becomes more useful.

   The research issue of CO_2 capture can be summarized as given in Table 3.2. Regarding screening of absorbents and energy performance evaluation, further expla-nations will be added below.

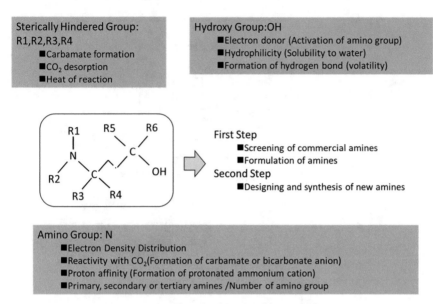

Fig. 3.6 Amine characteristics and R&D schematic

Table 3.2 Research issues in chemical absorption by amine-based solvents

I	Reduction of CO_2 capture cost – High-performance solvents with large CO_2 capacity – Novel absorbents with low-temperature regeneration – Innovative technology for chemical absorption
II	Development for practical application – Degradation and reclaiming – Corrosion – Environmental impact
III	Process optimization – Process simulation

(1) Screening and property measurements

Gas scrubbing test was an initial-stage attempt to figure out solution characteristics of CO_2 absorption and desorption [1]. 50 ml of an aqueous amine solution was put into a 250-ml glass scrubbing bottle, and a gas mixture (e.g., a simulated blast furnace gas: $CO_2/N_2 = 20/80$ vol.%) was supplied to the bottle. For the first 60 min, the bottle was placed in a water bath of 40 °C to investigate CO_2 absorption, and for the following 60 min, the bottle was placed in a water bath of 70 °C for a desorption test. The schematic diagram of a scrubbing system for a gas scrubbing test is shown in Fig. 3.7a.

Figure 3.7b also shows a typical result obtained from a gas scrubbing test. An amount of absorbed CO_2 in the aqueous amine solution, "CO_2 loading," was estimated from the measured CO_2 concentration of the outlet gas flow. The curve of

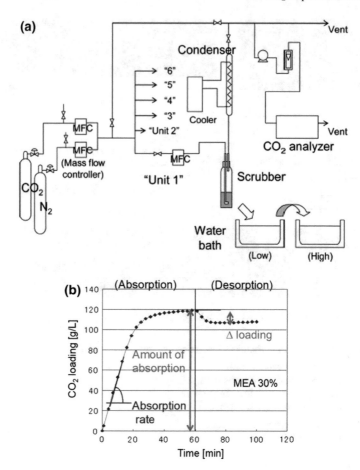

Fig. 3.7 Screening device and obtained solvent information

CO_2 loading with time showed CO_2 loading at 40 and 70 °C clearly. A gradient of the curve at a half point of the 60-min CO_2 loading was calculated as an absorption rate. This reference index was used to understand the relative behavior of an amine aqueous solution to develop candidate absorbents with better characteristics.

Figure 3.8 shows a schematic diagram of the experimental apparatus used for a vapor–liquid equilibrium (CO_2 solubility) measurement and a typical result. It consists of a 700-cm^3 crystal glass cylindrical reactor vessel, a steam saturator, a CO_2 analyzer, and so on. The gas mixture controlled to a specific CO_2 concentration was supplied to the reactor vessel. The equilibrium condition was determined when the CO_2 analyzer of the outlet gas indicated same CO_2 concentration of the inlet gas. To take the equilibrium data, the CO_2 concentrations in both gas and liquid phase were measured. The CO_2 partial pressure was derived from the temperature, total pressure, and the measured CO_2 concentration. For the liquid phase, a sample was

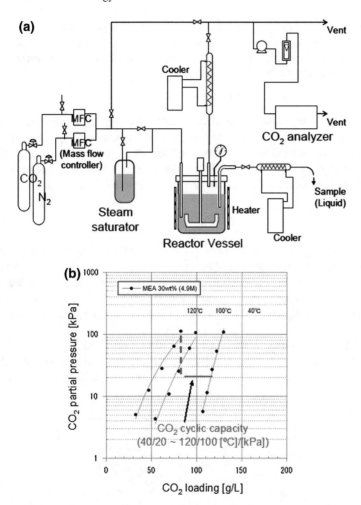

Fig. 3.8 Experimental apparatus and results of CO_2 solubility (vapor–liquid equilibrium measurement)

taken from the reactor vessel, and the amount of absorbed CO_2 in the sample was measured.

As well as the above-mentioned two solvent's information, absorption heat and reaction mechanism were important to develop new solvents. The heat of reaction for aqueous amine solutions was measured with a commercially available reaction calorimeter.

Also, ^{13}C NMR spectroscopy was a useful analysis equipment to analyze the CO_2-absorbed amine-based solvents and to determine the amounts of carbamate and bicarbonate anions produced by CO_2 absorption in a solvent. Regarding a solvent mainly absorbed CO_2 as bicarbonate anion, its CO_2 loading to one amine molecule becomes higher. Primary and secondary amines have a possibility to generate both carbamate and bicarbonate anions in the reaction with CO_2, but the ratio of the bicarbonate anion and carbamate anion is different for each amine compound. Tertiary amine absorbs CO_2 as only bicarbonate.

(2) Evaluation of energy performance

Thermal energy requirement is the critical factor for chemical absorption. It is defined as the reboiler heat duty divided by the amount of captured CO_2. It consists of the three energy consumptions (reaction, heat loss of solvent, and heat loss of steam at stripper top) as shown in Fig. 3.9, if the heat loss from CO_2-capture plant to atmosphere is ignored. This heat balance can be described as Eq. (3.1).

$$Q_T = Q_V + Q_H + Q_R \tag{3.1}$$

Thermal energy requirement can be estimated with heat and mass balance analysis of the system including stripper and rich/lean heat exchanger. In this analytical study, equilibrium stage model for stripper (regenerator) is firstly used to obtain the process data. Using the data, three heat consumptions defined in Eq. (3.1) can be calculated, respectively. Consequently, the thermal energy requirement can be obtained and evaluated.

If we need to analyze a chemical absorption process in detail, we can utilize a commercial process simulation software. We can evaluate a process engineering data such as stream compositions in the process, temperatures at process units, the

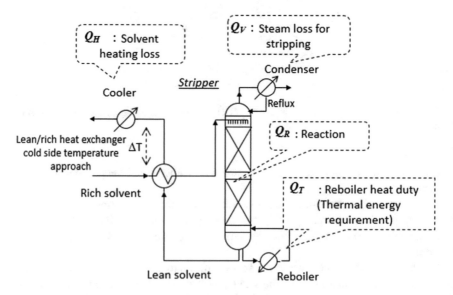

Fig. 3.9 Thermal energy requirement of chemical absorption

Fig. 3.10 Laboratory-scale test unit of chemical absorption for CO_2 capture, "CAT-LAB"

necessary reboiler heat duty, and so on. For example, a commercial process simulator, Aspen Plus®, is a well-known simulation tool.

In the R&D activities in RITE, new solvents selected as candidates with preferable features were tested in a laboratory-scale test unit, "CAT-LAB" (Fig. 3.10). Its configuration is a simple process flow consisting of one absorber column and one stripper column, like a process flow shown in Fig. 3.2. The plant capacity for CO_2 capture from a simulated blast furnace gas was up to about 5 kg-CO_2/d.

3.5.2 Results in Laboratory Stage

Both commercially available and synthesized amine absorbents were evaluated for their CO_2 loading capacity, absorption heat, and absorption rates, and a structure–performance relationship was discussed. Figure 3.11 shows a part of the results [2] and implies one of key features indicating the direction of our R&D. In the screening tests, we could see that there was the trade-off relation between heat of reaction and CO_2-absorption rate. We found that there was the trade-off relation for conven-

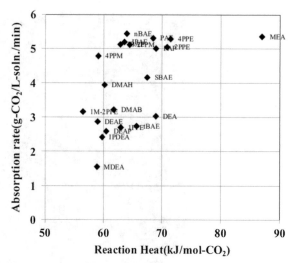

PAE: 2-(propylamino)ethanol, PAP: propylaminopropanol,
nBAE: 2-(butylamino)ethanol, IPAE: 2-(isopropylamino)ethnol,
tBAE:2-(tert-butylamino)ethnol, SBAE: (sec-butylamino)ethanol,
IBAE: 2-(isobutylamino)ethnol, 2PPM: 2-piperidinemethanol,
4PPM: 4-piperidinemethanol, 2PPE: 2-piperidineethanol,
4PPE: 4-piperidineethanol, DMAB: 4-(dimethylamino)butanol,
DMAH: 6-(dimethylamino)hexanol, DEAE: 2-diethylaminoethanol,
DEAP: 3-diethylamine-1-propanol, IPDEA: N-isopropyldiethanolamine,
1M-2PPE: 1-methyl-2-piperidineethanol, 1PPE: 1-piperidineethanol
MEA: monoethnolamine, DEA: diethanolamine, MDEA: N-methyldiethanolamine

Fig. 3.11 Relation between absorption heat and absorption rate [2]

tional absorbents and that candidate absorbents of high-performance solvents could be plotted outside of the trade-off relation. As a result of that, through screening various amine compounds, preferable ones were selected to formulate new solvents.

Table 3.3 and Fig. 3.12 show solvent features of a new high-performance solvent and a reference conventional solvent of MEA (30 wt% aqueous solution). Regarding the Solvent-A, although the absorption rate decreases slightly, the cyclic capacity becomes larger and absorption heat lowers significantly. Also, CO$_2$ solubility shown in Fig. 3.12 implies that the solvent can easily take a large amount of CO$_2$ into itself at an absorber and desorb most of CO$_2$ absorbed at a regenerator.

Figure 3.13 shows calculation and experiment results of an energy performance. The new solvent had a good performance in desorption characteristics, and their lean loadings were lower than that of MEA. Also, their capacities of CO$_2$ capture were larger as indicated in Fig. 3.12. Owing to the new solvents' advantages of low heat of reaction and large capacity of CO$_2$ capture, a thermal energy requirement was decreased in comparison with the MEA aqueous solution.

Table 3.3   Performance of a new solvent

Solvent	Absorption rate (g-CO_2/L/min)	Cyclic capacity (g-CO_2/L)	Absorption heat (kJ/mol-CO_2)
MEA 30 wt% aq.	5.5	62	87
Solvent-A	4.3	154	68

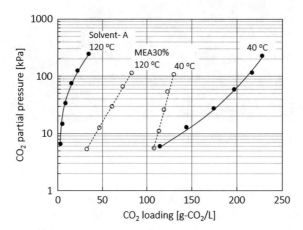

Fig. 3.12 CO_2 solubility of a new solvent

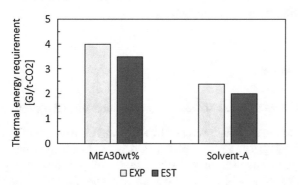

Fig. 3.13   Energy performance of a new solvent; EXP: experimental results of CAT-LAB, EST: calculation results by heat and mass balance analysis

3.5.3   Scale-up and Demonstration

The new solvents were tested in test plants installed at steelwork in order to verify a practical application. Figure 3.14 shows the test plant facilities. Regarding a bench plant (right side in Fig. 3.14), the plant capacity for CO_2 capture from BFG was up to about 1 t-CO_2/d. On the other hand, regarding process evaluation plant (left side), the plant capacity for CO_2 capture from BFG was up to about 30 t-CO_2/d. BFG of

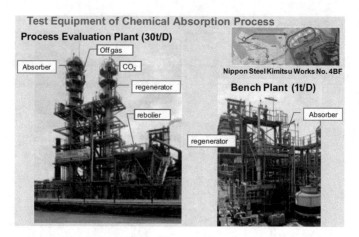

Fig. 3.14 Test equipment of chemical absorption process for CO$_2$ capture from blast furnace gas [18]

about 20 kPa-CO$_2$ in CO$_2$ partial pressure, which was supplied from a factory gas line, was blown into the absorber.

Moreover, the effects of variables, such as a BFG feed rate and a solvent circulation rate, a reboiler steam for CO$_2$ stripping and a stripper pressure were investigated in order to establish commercial technologies.

3.5.4 Outcomes

(1) Pilot plant test of a new solvent for post-combustion CO$_2$ capture

A new amine-based solvent developed by RITE was evaluated for CO$_2$ capture from the coal-fired flue gas of a power plant [3]. A pilot plant facility installed at a coal-fired power plant was used in this study. Its capacity to feed flue gas was about 2100 Nm^3/h (ca. 10 t-CO$_2$/d), which contained about 12% CO$_2$. The flue gas to the absorber was extracted at the downstream flow of a flue gas desulfurization process (FGD) and was supplied after another treatment by an additional FGD unit. Table 3.4 is a result of the pilot plant tests. After optimization of a reboiler steam feed and a liquid-to-gas ratio, 3.0 GJ/t-CO$_2$ of energy requirement was attained at 90% CO$_2$ recovery.

(2) Utilization of a new solvent at a commercial plant

One of the outstanding new absorbents developed by RITE was adopted for use in commercial CCS plants operated by Nippon Steel & Sumikin Engineering Co., Ltd. The first such CCS plant began operation in 2014 (Fig. 3.15) and the second CCS plant in 2018. The research results of RITE contribute to the practical application of CO$_2$-capture technology to different emission sources in the industry.

Table 3.4 Results of a pilot plant test in which the application of a new solvent to post-combustion CO_2 capture was evaluated [3]

Conditions of operation	
(Feed gas)	Feed rate: 2100 (Nm^3/h) Temperature: 35 (°C) CO_2 conc. (dry): 12 (%)
(Absorber)	Solvent temperature at inlet: 35 (°C)
(Stripper)	Pressure at top: 0.12 (MPaG)
	(Run 6)
L/G^a (kg-solvent/Nm^3)	2.2
Reboiler steam feed rate (GJ/h)	1.3
Results of operation	
(Absorber)	
CO_2 concentration of top gas (dry) (%)	1.3
Temperature at top (°C)	60.1
Temperature at bottom (°C)	37.5
(Stripper)	
Temperature at top (°C)	100.0
Temperature at bottom (°C)	124.6
(Performance)	
CO_2 recovery (%)	91
Energy requirement (GJ/t-CO_2)	3.0

^{a}L/G: Liquid-to-gas ratio

3.6 Exploring New Research Area

R&D on chemical absorption at RITE has been shown in Sect. 3.5, but our activities are advancing to new objects taking advantage of knowledge about accumulated absorbent here. Technology called "high-pressure regenerative chemical absorption method" is actively addressing RITE. This section will briefly describe technologies of high-pressure regenerative chemical absorbent.

3.6.1 Chemical Absorption with High-Pressure Regeneration

Dissolution of the gas into the liquid is easier as the partial pressure of the target gas is higher. This is convincing if you remember physical absorption. Therefore, targets, under conditions where the partial pressure of the target gas is high, are put to practical use in industry, for instance, purification of natural gas, removal of carbon

Fig. 3.15 Snapshot of the first commercial plant for supplying CO$_2$ as industrial gas. The construction site is Nippon Steel & Sumitomo Metal Corporation. The photo is provided by Nippon Steel & Sumikin Engineering Co., Ltd.

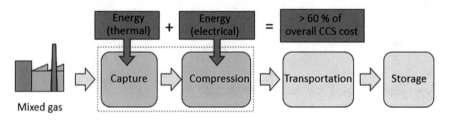

Fig. 3.16 CO$_2$ capture and storage (CCS)

dioxide from gas mixture of hydrogen and carbon dioxide after steam reforming of hydrocarbon, and the like.

The general process flow is as shown in Sect. 3.2, and it is necessary to drive out by heat in order to sufficiently lower the amount of dissolved CO$_2$ in the lean solution.

By the way, some researchers propose that because CCS is a system combining several unit operations as shown in Fig. 3.16, not a separate examination of only the unit process, but a comprehensive examination of adjacent processes should be focused on to build a more optimal system. We proposed a technical study to reduce energy consumption by integrating recovery process and compression process.

High-pressure regenerative chemical absorption concept was presented by the University of Texas [5], JGC [17], and RITE [19]. RITE's research will be discussed in the next section.

3.6.2 Novel Absorbents for High-Pressure Regeneration

RITE has advanced the development of chemical absorbents for high-pressure gas mixtures containing CO_2, obtaining excellent CO_2 absorption and desorption performance. The purpose of this R&D is to develop highly efficient absorbents that enable CO_2 desorption while maintaining the high CO_2 partial pressure of the initial gas mixture (Fig. 3.17). The concept of this new CO_2-capture technology is named as high-pressure regenerable absorbents. Using these novel absorbents, the energy consumption during the CO_2 compression process following capture is significantly reduced owing to the high pressure of the captured CO_2.

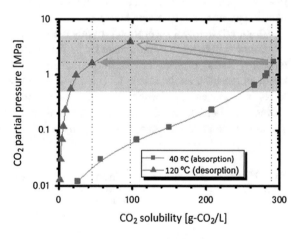

Fig. 3.17 Novel chemical solvent and process for CO_2 capture from pressurized gas mixture

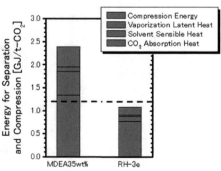

Fig. 3.18 Overall energy requirement from CO_2 capture to compression

Several new solvent systems designed previously have demonstrated high CO_2 recovery levels in conjunction with superior CO_2 absorption and desorption rates. These capabilities are in addition to a relatively low heat of reaction under high-pressure conditions above 1 MPa (Fig. 3.18). The total energy consumption rate for CO_2 separation and capture when using this process, including the energy required for compression, has been estimated to be less than 1.1 GJ/t-CO_2 (absorption: 1.6 MPa-CO_2, desorption: 4.0 MPa-CO_2).

3.7   Conclusion

Absorption is one of gas separation process technologies. In particular, a method of separating a target gas from a gas mixture using a chemical absorbent becomes a hot R&D topic. The application of the chemical absorption to CO_2 capture of CCS is more advanced than other gas separation technologies, because of higher technology readiness level of the chemical absorption. Also, we could say that a coal-fired power plant with CO_2 capture is in a ready state. In fact, CCS-EOR using post-combustion has already been put to practical use.

Although chemical absorption is highly mature, it still has the problem that energy requirement is so high to implement CCS. Therefore, new absorbents and processes of energy-saving process have been researched and developed. RITE has developed novel amine-based solvents for CO_2 capture from blast furnace gas and succeeded to attain lower energy consumption than conventional solvent. Moreover, the new solvent has been used in industry as a commercial chemical absorbent with the support of the private engineering company.

Fig. 3.19 Challenge for innovation

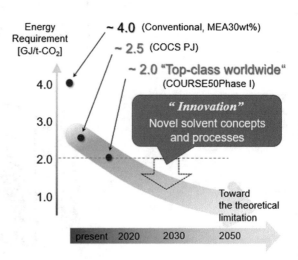

In addition, RITE has developed high-pressure regenerative chemical absorbing liquid by utilizing knowledge of research and development of chemical absorbent using amine compound and succeeded in consumption energy of 1.1 GJ/t-CO_2.

The last figure of Fig. 3.19 shows the challenge of chemical absorption for CCS. It is not easy to promote practical applications of CCS as a technological countermeasure for global warming along the CO_2 mitigation scenario. We need a strong driving force. Therefore, further efforts to research and development of chemical absorption are required to obtain innovative technologies that further pursue low energy consumption and low cost.

Acknowledgements The R&D of RITE was carried out under COCS project, which was funded by the Ministry of Economy, Trade and Industry (METI), Japan, and also under the COURSE 50 projects, which is entrusted the New Energy and Industrial Technology Development Organization (NEDO).

References

1. Chowdhury FA, Yamada H, Higashii T, Goto K, Onoda M (2013) CO_2 capture by tertiary amine absorbents: a performance comparison study. Ind Eng Chem Res 52:8323–8331
2. Chowdhury FA, Yamada H, Higashii T, Matsuzaki Y, Kazama S (2013) Synthesis and characterization of new absorbents for CO_2 capture. Energy Procedia 37:265–272
3. Goto K, Kodama S, Higashii T, Kitamura H (2014) Evaluation of amine-based solvent for post-combustion capture of carbon dioxide. J Chem Eng Japan 47(8):663–665
4. Liang Z, Rongwong W, Liu H, Fu K, Gao H, Cao F, Zhang R, Sema T, Henni A, Sumon K, Nath D, Gelowitz D, Srisang W, Saiwan C, Benamor A, Al-Marri M, Shi H, Supap T, Olson W, Idem R, Tontiwachwuthikul P (2015) Recent progress and new developments in post-combustion carbon-capture technology with amine based solvents. Int J Greenh Gas Control 40:26–54
5. Lin Y-J, Rochelle GT (2014) Optimization of advanced flash stripper for CO_2 capture using piperazine. Energy Procedia 63:1504–1513
6. Mondal MK, Balsora HK, Varshney P (2012) Progress and trends in CO_2 capture/separation technologies: a review. Energy 46:431–441
7. Idem R, Supap T, Shi H, Gelowitz D, Ball M, Campbell C, Tontiwachwuthikul P (2015) Practical experience in post-combustion CO_2 capture using reactive solvents in large pilot and demonstration plants. Int J Greenh Gas Control 40:6–25
8. IEA (2014) Energy technology perspective
9. IEA (2016) World energy outlook
10. IEA (2018) CO_2 emissions from fuel combustion
11. IEAGHG (2013a) Overview of the current state and development of CO_2 capture technologies in the ironmaking process. Report: 2013/TR3
12. IEAGHG (2013b) Development of CCS in the cement industry. Report: 2013/19
13. IPCC (2005) Special report of carbon dioxide capture and storage. Cambridge University Press
14. Rayer AV, Sumon KZ, Sema T, Henni A, Idem R, Tontiwachwuthikul P (2012) Part 5c: solvent chemistry: solubility of CO_2 in reactive solvents for post-combustion CO_2. Carbon Manag 3:467–484
15. Shi H, Liang Z, Sema T, Naami A, Usubharatana P, Idem R, Saiwan C, Tontiwachwuthikul P (2012) Part 5a: solvent chemistry: NMR analysis and studies for amine-CO_2-H_2O systems with vapor-liquid equilibrium modeling for CO_2 capture processes. Carbon Manag 3:185–200
16. Tan Y, Nookuea W, Li H, Thorin H, Yan J (2016) Property impacts on carbon capture and storage (CCS) processes: a review. Energy Convers Manag 118:204–222

17. Tanaka K, Fujimura Y, Komi T, Katz T, Spuhl O, Contreras E (2013) Demonstration test result of High Pressure Acid gas Capture Technology (HiPACT). Energy Procedia 37:461–476
18. Tonomura S (2013) Outline of COURSE50. Energy Procedia 37:7160–7167
19. Yamamoto S, Machida H, Fujioka Y, Higashii T (2013) Development of chemical CO$_2$ solvent for high pressure CO$_2$ capture. Energy Procedia 37:505–517
20. IEAGHG (2014) Assessment of emerging CO$_2$ capture technologies and their potential to reduce costs, 2014/TR4

Chapter 4
CO$_2$ Capture with Adsorbents

Abstract The adsorption method for CO$_2$ capture is described in this chapter. The several types of absorbents including amine-based adsorbent and other ones are reviewed. Then, the advanced absorbents, i.e., "solid sorbents" developed in a project in RITE are described. The solid sorbents are composed of amine absorbents for chemical absorption and porous materials. They have similar CO$_2$-adsorption characters with liquid amine absorbents. Furthermore, they make it possible to significantly reduce the energy consumed as sensible heat and evaporative latent heat in the regeneration process. Novel amines synthesized in this project have been employed for their developed solid sorbents to successfully fabricate innovative, high-performance solid sorbents capable of low-temperature regeneration with high-adsorption capacities. Using the novel amines obtained by the improved-synthesis method for a large scale, the moving-bed bench-scale plant test was recently started.

Keywords Adsorption · Zeolites · Solid sorbent · Mesoporous materials · Amine

4.1 Outline of Adsorption Separation Method

4.1.1 Introduction

The CO$_2$-adsorptive separation method has been put to practical use to date, and it is also being studied as a method for separating and recovering CO$_2$ from a large-scale source. On the other hand, it is also necessary to further reduce the separation energy and make the apparatus compact. It is a great merit that startup stop and operation of the equipment is easy and operation of waste liquid is unnecessary, and if a chemically stable high-performance adsorbent is developed, substantial cost reduction and energy saving are possible. In this chapter, in addition to overview and recent trends of adsorption method, we will introduce the approach to R&D on new energy-saving-type CO$_2$-adsorptive separation method currently underway at RITE.

© The Author(s), under exclusive license to Springer Nature Switzerland AG 2019   45
S. Nakao et al., *Advanced CO$_2$ Capture Technologies*,
SpringerBriefs in Energy, https://doi.org/10.1007/978-3-030-18858-0_4

4.1.2   Adsorption Separation Method

The adsorptive separation method refers to a method of adsorbing a specific component in a gas or liquid onto a porous solid (adsorbent) and separating, concentrating, removing, and collecting the components.

Absorption method and adsorption method essentially are separation methods based on the same principle. CO_2 can be separated from CO_2-containing gas by contact with liquid absorbent or solid adsorbent having an affinity for CO_2. Transfer the substrate reacted with the recovered CO_2 to a different container and change the conditions such as heating or decompression in the container to release and collect CO_2 and regenerate the substrate. In the case of scrubbing absorbent, the absorbent (liquid) moves between the two containers and repeats absorption/regeneration. In the case of the solid adsorbent, in general, the adsorbent does not move in the container. As shown in Fig. 4.1, the external environment is changing within the same container. (It is called a fixed-bed system; there is also a moving-bed system in which the adsorbent is moved.) Since solid adsorbent is difficult to move containers, it is disadvantageous compared with the chemical absorbent in case of performing heating regeneration. On the other hand, there is no need to consider the steam loss which is a problem with the chemical absorbent; there is a possibility of energy consumption reduction.

Both absorbing liquid and solid adsorbent are always required to be supplied to compensate for performance deterioration and dissipation. Currently, new processes using new absorbing liquids and solid adsorbents have been developed, but the common problem among them is that they must cope with large amounts of CO_2

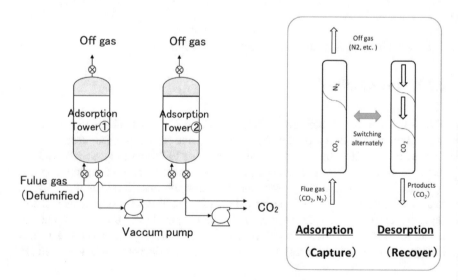

Fig. 4.1 Principle of adsorption separation method (PSA)

Table 4.1 Physical adsorption and chemisorption

Type	Physical adsorption	Chemical adsorption
Principle	van der Waals	Chemical bond (electron transfer)
Temperature	Large amount of adsorption at low temperature	It occurs at a relatively high temperature
Adsorbate	Nonselective	Selective
Heat of adsorption	Small (8–20 kJ mol^{-1}) (Equivalent to condensation heat of adsorbate)	Large (40–800 kJ mol^{-1}) (Equivalent to heat of reaction)
Reversibility	Reversibility	There may be cases of irreversibility
Adsorption rate	Fast	Slow (requiring activation energy)

treatment. In addition, when absorption liquid/adsorbent is expensive, it is necessary to consider the price of absorbing solution/adsorbent and processing cost of degradation products.

The adsorption phenomenon has weak physical adsorption by van der Waals force and strong chemical adsorption with the chemical bond. In the former physical adsorption, mainly zeolite or activated carbon is used as CO_2 adsorbent. In the separation method using physical adsorption, the adsorbent can be repeatedly used by desorbing and regenerating adsorbed substances by changing temperature and pressure during adsorption.

On the other hand, as an adsorbent for chemical adsorption, amine-supported inorganic porous material, hydrotalcite, calcium oxide, lithium silicate, and the like are used [4]. Chemical adsorbents are often used at higher temperatures than physical adsorbents. However, adsorbents used for special uses (such as manned space activities) such as lithium hydroxide and silver oxide are difficult to regenerate, and sometimes they are used at low temperatures and thrown away (Tables 4.1 and 4.2).

4.1.3 CO₂ Separation and Recovery Technology by Physical Adsorption Method

In CO_2 separation and recovery by physical adsorption, CO_2 is adsorbed to an adsorbent that selectively adsorbs CO_2 rather than other gases due to relatively weak interactions caused by the van der Waals forces. Thereafter, the adsorbed CO_2 is desorbed by decompression or heating to separate and recover the concentrated CO_2.

There are two kinds of methods for desorbing CO_2: pressure swing adsorption (PSA) method using pressure difference and thermal swing adsorption (TSA) using temperature difference. That is, after separating a specific gas adsorbed on the adsorbent from a component not adsorbed, the former is desorbed by depressurization and used mainly for the physical adsorption method, whereas the latter is desorbed

Table 4.2 Classification of CO_2 adsorbent/sorbent

Type	Adsorbent	Application
Physical adsorbent	Zeolites (5A, 13X)	Gas purification Space station (ISS) R&D stage (CO_2 capture from blast furnace)
	Activated carbon	Gas purification
	MOFs	R&D stage (for high-pressure CO_2 separation)
Chemical sorbent	Amine-modified-oxides/carbon/MOFs	R&D stage (post-combustion capture)
	Hydrotalcites	R&D stage (sorption enhanced reaction, etc.)
	Potassium-doped carbon Nitrogen-doped carbon	R&D stage (post-combustion capture)
	Calcium oxides Lithium silicate/zirconate	R&D stage (chemical looping)
	LiOH, AgO (MetOx)	Life support system of space suit

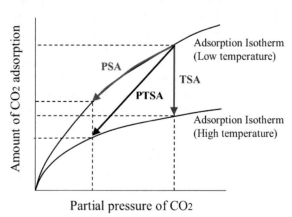

Fig. 4.2 Principle of desorption of PSA and TSA method

by heating. It is mainly used for chemisorption. Since the cycle time can be short-ened, the PSA method is widely used for practical purposes such as oxygen–nitrogen separation and hydrogen refining.

In addition, as shown in Fig. 4.2, the pressure and temperature swing adsorption method (PTSA method) in which PSA and TSA are combined has also been studied. By heating the adsorbent at the time of desorption, the PTSA method can increase the regeneration capacity of the adsorbent and also reduce the power required for decompression (mainly vacuum pump power). If it is possible to use unused energy, etc., in the thermal power plant as a heat source, it is expected to be an economically superior system.

(1) Zeolites

The most representative adsorbent used for physical adsorption is zeolite. Zeolite is a crystalline aluminosilicate, and gas adsorption to zeolite is mainly due to the interaction between localized electrostatic field caused by cations bonded to $(AlO_4)^-$ in the framework structure and polarity of molecules by physical adsorption. Therefore, it selectively adsorbs polar molecules and polarized molecules having dipole and quadrupole moments. Further, zeolite has three-dimensional uniform pores of several Å levels close to the molecular diameter of the gas (Fig. 4.3).

These pores are composed of oxygen four- to fourteen-membered rings and show the molecular sieves' effect which can control the diffusion of molecules into the pores depending on the size of these oxygen rings. Therefore, for the purpose of gas separation purification, the adsorption molecule size must be smaller than the zeolite pore size.

Generally, it is said that the selectivity of polar gas considered to be an impurity present in various raw material gases is as follows.

$$H_2O > C_2H_5SH > CH_3SH > NH_3 > H_2S > SO_2 > COS > CO_2$$

As shown in the above ranking, the most selective molecule for general zeolite is water and its adsorption capacity is also extremely large, so it is often used as a hygroscopic material for drying gas and liquid in the petrochemical process.

Therefore, when zeolite is used as an adsorbent for separating and recovering CO_2, it is necessary to remove water beforehand by a pre-treatment process.

(2) Metal-organic frameworks (MOFs)

A new class of porous materials constructed from metal ions and organic ligands, known as metal–organic frameworks (MOFs) or porous coordination polymers (PCPs), has attracted much attention in the field of material chemistry. There are many reports on MOF, so please refer to that [22, 23, 27, 33]. As shown in Table 4.3, many MOFs have been reported to outperform traditional porous solids such as zeolites in terms of CO_2-capture capabilities [13, 31]. Unfortunately, a significant issue regarding the application of MOFs in industrial CO_2 capture is H_2O

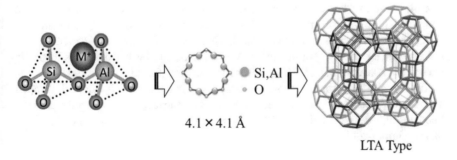

4.1 × 4.1 Å

LTA Type

Fig. 4.3 Framework structure of zeolite (LTA type)

Table 4.3 High-pressure
CO$_2$-adsorption capacities in
selected MOFs at 313 K

	Pressure (MPa)	Capacity (mol/kg)
MOF-177	0.1, 1.6, 3.9	0.9, 16.2, 33.8
BeBTB	0.1, 1.6, 3.9	0.9, 16.4, 30.2
CoBDP	0.1, 1.6, 4.0	0.3, 11.7, 16.6
CuBTTri	0.1, 1.6, 3.9	2.2, 13.1, 17.0
Mg$_2$(dobdc)	0.1, 1.6, 3.5	8.0, 14.3, 15.1
HKUST-1	0.4, 1.5, 3.9	8.4, 12.7, 13.7

sensitivity, which leads to decreased MOF performance or collapse of framework [3, 24, 36].

(3)   Issues of the current physical adsorption separation method

In the conventional physical adsorption method, since the CO$_2$-adsorption amount of the zeolite used as the adsorbent decreases significantly in the presence of water vapor, water vapor in the exhaust gas is separated and removed as a pre-treatment, and CO$_2$ is adsorbed and separated in the subsequent stage. When zeolite is used as an adsorbent for CO$_2$, it must be used at a dew point of -30 to -60 °C. In this case, about 30% of CO$_2$ separation recovery energy is consumed for dehumidification. Zeolitic adsorbents and activated carbon are used or studied in conventional CO$_2$ physical adsorption methods.

Among them, zeolite 13X is regarded as superior in CO$_2$-adsorption capacity [1, 21]. As shown in Fig. 4.4, the X-type zeolite exhibits CO$_2$-adsorption characteristics of the Langmuir type, and a high-adsorption amount of CO$_2$ can be obtained at a low CO$_2$ partial pressure of about 10–15 kPa corresponding to the CO$_2$ concentration of the exhaust gas of the thermal power plant. However, in the desorption process, pressure reduction (PSA) or heating (TSA) operation by a vacuum pump is required, and a large amount of energy is required.

The adsorption amount of CO$_2$ is as large as 5 mol kg^{-1} at 313 K, but it cannot be said that the adsorption amount reaches a sufficient amount at 333–373 K corresponding to the exhaust gas temperature [28]. If an increase in adsorption amount is achieved in this temperature range, it is possible to make the apparatus compact. Moreover, in addition to energy saving by the adsorption method, compactification of the apparatus is also an important subject. Therefore, CO$_2$ separation can be carried out even under water vapor coexistence conditions, and further materials that may exceed the conventional 13X CO$_2$-adsorption amount should be examined.

If a chemical adsorbent capable of adsorptive separation of CO$_2$ can be developed even in the presence of water vapor, a water vapor removal process as a pre-treatment becomes unnecessary, and reduction of CO$_2$ separation recovery energy and process can be simplified.

Fig. 4.4 CO_2-adsorption isotherm of zeolite 13X (313 K). □: Under dry conditions, ■: under wet conditions

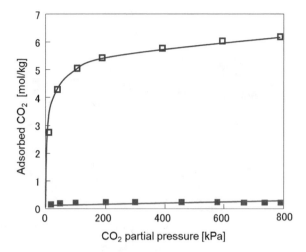

4.2 CO₂ Separation and Recovery Technology by Chemical Adsorption Method Using Mesoporous Material

Organically functionalized mesoporous silica has attracted considerable attention because of the wide range of applications as adsorbents and catalysts [2, 9, 18, 20, 44, 45]. Because mesoporous silica has uniform and large pores, and a relatively high surface area, a large number of active sites or adsorption sites can be uniformly introduced on the pore walls via surface modification with organosilane molecules.

The mesoporous materials are substances located between microporous materials such as zeolites and macroporous materials such as porous glass and have pore diameter of about 2–50 nm. Since it is synthesized from two-dimensional or three-dimensional aggregates composed of inorganic chemical species such as surfactant and hydroxide, unlike ordinary materials having mesopores such as silica gel, the pore size distribution is uniform and very narrow.

In the early 1990s, mesoporous silica was separately synthesized using organic templates by Kuroda's group [19] and researchers of Mobil [26]. They have a regular pore structure, a uniform pore size of the mesoporous region, and a large specific surface area and a pore volume. Due to its interesting properties, mesoporous silica attracts attention and development of a novel synthesis method has been carried out and application to a support material such as a catalyst or adsorbent targeting a molecule larger than the pores of zeolite has been studied [5].

Since mesoporous silica has large pore size and pore volume, large molecules can be introduced into pores by chemical bonding (Fig. 4.5). As an advantage of utilizing such a solid absorbent, prevention of evaporation of amine and ease of handling can be expected.

In addition, mesoporous materials can synthesize not only silica but also substances having various compositions and different pore structures such as sulfide,

Synthesis of mesoporous materials

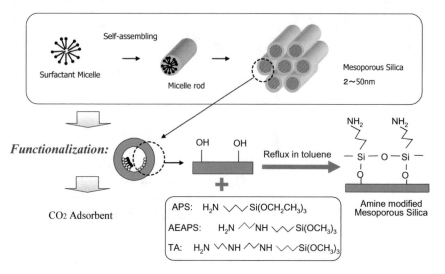

Fig. 4.5 Surface amine modification of mesoporous silica

phosphate, carbon, and the like. Further, it is possible to hybridize with an organic compound by modifying the surface with various functional groups. Therefore, it is possible to design various materials, and it is also expected to create materials that exhibit excellent gas adsorption/desorption performance.

Therefore, many researches on CO$_2$ adsorption by various amine-modified mesoporous materials have been reported so far [2, 6, 11, 14–18, 30, 42, 44, 45].

Amine-functionalized porous materials for CO$_2$ capture can be classified into three types of materials, including amine-impregnated (Class 1), amine-grafted (Class 2), and in situ polymerized amine-grafted (Class 3) materials [29]. Silicas are mainly used for the support material. The pore size of the support materials, the molecular structures of the amines, the amine composition, and the adsorption temperature were found to play important roles and greatly affect the CO$_2$-adsorption performance.

Hiyoshi et al. [16] evaluated the CO$_2$-adsorption properties of various aminosilane-grafted SBA-15 mesoporous silica (Class 2). SBA-15 exhibited high CO$_2$ uptake, and the CO$_2$-adsorption capacity increased with the surface density of amines. (This will be described in detail in Sect. 4.4.1.) However, introduction of amine by grafting was limited by the numbers of surface OH groups.

On the other hand, the wet impregnation (Class 1) is a simple preparation method, and a larger number of amines can be introduced into the pores of the support materials, leading to a higher-CO$_2$-adsorption capacity compared to that of grafting methods [48]. For the preparation of sorbents with high-CO$_2$-adsorption performance by the impregnation method, polyamines such as tetraethylenepentamine (TEPA) [10, 39, 50], pentaethylenetetramine (PEHA) [35], diethylenetriamine (DETA) [43], and polyethylenimine (PEI) [38, 41] have often been used to modify the supports.

Structurally ordered porous silica materials with high surface areas and large pore diameters/volumes such as SBA-12 [9], SBA-15 [16], MCM-41 [7, 8, 49], SBA16 [25], MSU [40], and MSF [47] are often selected to load a larger number of amines into the pore channels.

To date, several new high-performance CO$_2$ sorbents were developed by the impregnation of TEPA and amino alcohols into mesostructured silica materials [6]. The CO$_2$-adsorption capacity of amine-impregnated silica can be improved by blending organic compounds containing hydroxyl groups. Dao et al. reported that TEPA40-DEA30/MSU-F sorbent, which is based on the largest pore silica (MSU-F), was impregnated with TEPA (40 wt%) and DEA (30 wt%) and it had the largest CO$_2$-adsorption capacities of 5.91 mmol/g under a pressure of 100 kPa at 323 K.

On the other hand, Quyen et al. recently reported that adding imidazoles containing electron-donating groups to TEPA synergistically improved the CO$_2$-adsorption capacity, amine efficiency, working capacity, and regeneration energy requirement of solid sorbents. It was reported that positive interactions between imidazoles and the amino group may have been caused by the proton-acceptor ability of imidazoles and improvements in the diffusion of protons and CO$_2$ to subsurface reactive sites. A mesostructured cellular silica foam impregnated with 30% 4-methylimidazole and 40% TEPA exhibited a high-adsorption capacity (5.88 mmol/g) under conditions of 323 K and 100 kPa CO$_2$ [32].

4.3 Recent Development of CO$_2$-Adsorption Separation Method

In Japan, adsorptive separation technology is being studied as part of a project to significantly reduce CO$_2$ emanating from blast furnaces. Meanwhile, in the USA, adsorption separation by a solid absorbent is being studied as a technology for recovering CO$_2$ from coal-fired power plants.

The Japanese project CO$_2$ ultimate reduction in steelmaking process by innovative technologies for cool earth 50 (COURSE50) funded by New Energy and Industrial Technology Development Organization (NEDO) is led by the Japan Iron and Steel Federation (JISF) in collaboration with six major steel industries and related companies, and has led to the development of environmentally benign steelmaking technologies, including CO$_2$ capture. JFE Steel Corp. built a 3-ton-CO$_2$/day scale pressure swing adsorption (PSA) apparatus, called ASCOA-3, to evaluate the CO$_2$-capture performance from blast furnace gas. They demonstrated that the purity and recovery of CO$_2$ by zeolites was >90 and 80%, respectively [34]. Because zeolites adsorb H$_2$O vapor more strongly than CO$_2$, post-combustion CO$_2$ capture in the presence of H$_2$O vapor by this adsorption method requires a dehumidification step that consumes approximately 30% of the total CO$_2$-separation energy. If solid sorbents with H$_2$O tolerance are obtained, the capture devices can be compacted and energy can be saved by eliminating the dehumidification step.

Recently, amine-modified solid sorbents such as polyethyleneimine-loaded silica have fascinated many researchers with their adsorption characteristics such as H$_2$O tolerance. In the USA, the National Energy Technology Laboratory (NETL) developed amine-modified solid sorbents from liquid amines and clay minerals [37]. The clay-amine sorbents, capable of capturing CO$_2$ in the presence of H$_2$O vapor at 30–60 °C and regeneration at temperatures around 80–100 °C, were awarded one of the R&D Magazine's "R&D 100" awards in 2009. According to the preliminary system analysis conducted at the NETL, the regeneration energy required for clay-amine sorbents was 2.5 times lower than that for conventional aqueous amine scrubbing. In addition, $15 million per year was saved in a power plant of 550 MW and was estimated to potentially amount to $450 million over the 30-year lifespan. Research Triangle Institute (RTI), in collaboration with its partners, has developed an advanced solid-sorbent-based CO$_2$-capture process that can substantially reduce the energy requirement and costs associated with capturing CO$_2$ from coal-fired power plants. A promising sorbent combined with RTI's circulating fluidized moving-bed reactors (FMBRs) has been evaluated on the laboratory scale on the actual coal-fired flue gas to demonstrate the feasibility of the concept.

We conducted a project funded by the Ministry of Economy, Trade and Industry (METI) in 2010, in which the research objectives included the development of amine-modified solid sorbents for more efficient CO$_2$ capture and the establishment of evaluation standards of CO$_2$-capture systems. We are in the process of fabricating amine-modified solid sorbents applicable for CO$_2$ capture from coal-fired power plants with a goal of 1.5 GJ/t-CO$_2$. Herein, our research on amine-modified solid sorbents is reviewed.

4.4 New CO$_2$-Adsorption Separation Technology Development at RITE

In the following, we will introduce our approach to research and development of energy-saving-type new CO$_2$-adsorptive separation method we are working on.

4.4.1 Amine-Grafted Mesoporous Silica

RITE has developed a steam-resistant solid sorbent by grafting amino groups onto the surface of mesoporous silica. Mesoporous materials exhibit larger pore diameters (2–50 nm) and greater pore volumes than the microporous materials that are often used for gas capture and separation, allowing the size and amount of grafted amines to be tunable over a wide range (Fig. 4.6). This solid sorbent is able to simplify the CO$_2$-capture process and reduce the associated energy demands due to its steam

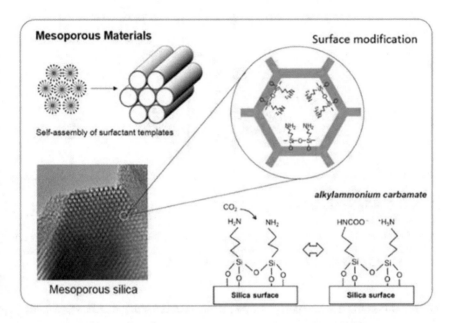

Fig. 4.6 Adsorption mechanism of CO$_2$ on aminosilane-modified mesoporous silica

resistance. Furthermore, reduced amine losses and improved ease of handling are also expected by immobilization of the amine on the solid support.

Solid supports may have a variety of chemical compositions, pore morphologies, and surface functional groups. The associated range of designs thus allows the synthesis of novel materials for CO$_2$ capture with high-adsorption/desorption performance. RITE has found that a solid sorbent with densely grafted amine groups exhibits high-CO$_2$-adsorption performance under wet conditions. Mesoporous silica SBA-15 was grafted using various aminosilanes: 3-aminopropyltriethoxysilane (H$_2$NCH$_2$CH$_2$CH$_2$-Si(OCH$_2$CH$_3$)$_3$, abbreviated as APS), N-(2-aminoethyl)-3-aminopropyltrimethoxysilane (H$_2$NCH$_2$-CH$_2$NHCH$_2$CH$_2$CH$_2$Si(OCH$_3$)$_3$, AEAPS), and (3-trimethoxysilylpropyl)diethylenetriamine (H$_2$-NCH$_2$CH$_2$NHCH$_2$CH$_2$NHCH$_2$CH$_2$CH$_2$Si(OCH$_3$)$_3$, TA).

The relationship between the amine content of the adsorbents and the CO$_2$-adsorption capacities is plotted in Fig. 4.7a. The amount of adsorbed CO$_2$ with APS-, AEAPS-, and TA-modified SBA-15 [14, 16] increased with increased amine content. Notably, the adsorption capacities were not in proportion to the amine content; instead, it increased exponentially. Figure 4.7b shows the amine efficiency defined by Eq. (4.1) for APS-, AEAPS-, and TA-modified SBA-15 as a function of amine surface density.

Amine efficiency $[-]$ = adsorbed CO$_2$ $[\text{mmol/g}]$/amine content $[\text{mmol/g}]$ (4.1)

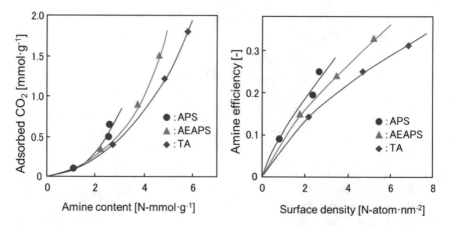

Fig. 4.7 Relationship between amine content and CO_2-adsorption capacity (**a**) and that between amine density and amine efficiency (**b**) under 15-kPa CO_2 and 12-kPa H_2O with a N_2 balance at 333 K. (●) APS-, (▲) AEAPS- and (■) TA-modified SBA-15

The amine efficiency for these adsorbents increased with increased surface density. Thus, the CO_2-adsorption site seems to be the densely anchored amines, rather than the isolated amines. Based on our IR results, strong interactions between amines and silanols, particularly at low-amine surface densities, may be effective for CO_2 adsorption.

This is possible because the formation of carbamates (R_2NCOO^-) by reaction of amine groups and CO_2 is not affected by moisture above 333 K. As an example, the absorbent triamine-grafted MSU-H (TA/MSUH) shows CO_2-adsorption performance under wet conditions similar to that of Zeolite 13X under dry conditions (Fig. 4.8) [17]. This property could allow omission of the dehumidification process typically required prior to CO_2 absorption, resulting in a reduction in the size of the process equipment. Recently, we have also investigated the application of this material to life support in space.

4.4.2 Amine-Impregnated Solid Sorbent

The solid-sorbent technology holds promise for the advantage of a lower-expected heat duty for regeneration processes by using porous-material-supported amines, exhibiting similar CO_2-sorption characteristics to those of amine-based solvents (Fig. 4.9).

As shown above, the amine-grafted/impregnated mesoporous material exhibits excellent CO_2-adsorption performance. Ordered mesoporous silica supports with large pore volumes, large pore sizes, and good pore interconnections tend to improve the CO_2-capture capacity of the sorbents. For this reason, silica mesostructured mate-

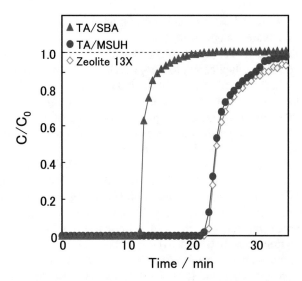

Fig. 4.8 CO$_2$ breakthrough curves on amine-modified mesoporous silica and zeolite 13X gas composition: CO$_2$ (15%)–H$_2$O (12%)–N$_2$ (TA/MSUH, TA/SBA) CO$_2$ (15%)–N$_2$ (Zeolite 13X); temperature: 313 K

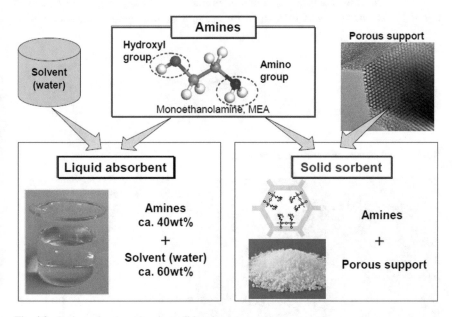

Fig. 4.9 Amine solvents and amine solid sorbents

rials such as MSU and MSF with large uniform pores make them promising candidates as supports of solid amine sorbents.

However, these materials (aminosilane reagent and ordered mesoporous silicas) will not be suitable for application to "large-scale stationary sources." In order to apply it to large-scale sources such as power plants, it is indispensable to be able to synthesize a large amount of adsorbent at low cost. Therefore, based on the above knowledge of the amine-modified mesoporous materials, we are also developing a solid-sorbent material in which amine is wet impregnated into a practical (= commercially available and inexpensive) porous support.

RITE developed the solid sorbents in the project for the advancement of CO$_2$-capture technologies consigned by METI from 2010 to 2014. Based on the established relationship between amine structures and their CO$_2$-desorption performances by simulation studies, RITE successfully fabricated innovative solid sorbents with high performance using newly synthesized amine in terms of low-temperature regeneration and sorption capacity (Fig. 4.10) [12, 46].

Then, we evaluated the processes of RITE solid sorbent using a lab-scale adsorption/regeneration test apparatus (Fig. 4.11) and found that the steam desorption process under vacuum, namely steam-aided vacuum swing adsorption (SA-VSA), remarkably enhanced the CO$_2$ recovery, compared with the normal VSA process.

A three-column fixed-bed system was employed for CO$_2$ capture as shown in Fig. 4.11. Desorption was performed by vacuum or low-temperature steam (i.e.,

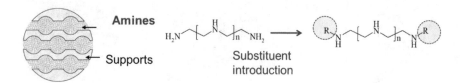

Fig. 4.10 Substituent introduction for polyamine

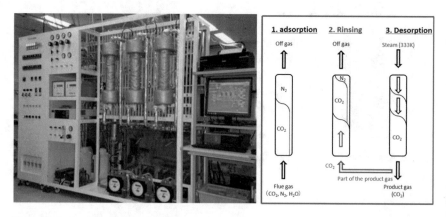

Fig. 4.11 Lab-scale apparatus for CO$_2$-capture test (left) and CO$_2$ recovery process (right)

steam-aided vacuum). The separation process consisted of three steps (i.e., adsorption, rinsing, and desorption). In the adsorption process, simulated flue gas was supplied to a column and CO_2 was captured. In the following rinsing process, a part of the recovered high-purity CO_2 was used to force out impurities such as N_2 from interparticle voids. Finally, in the VSA desorption process, the column was evacuated using a vacuum pump. For SA-VSA operation, the column was evacuated using a vacuum pump, and steam generated in a vaporizer was supplied to the column.

Further, we carried out the optimization of SA-VSA process. As a result, we demonstrated that RITE solid sorbent recovered CO_2 with high purity (>99%) and high recovery yield (>90%) from a simulated flue gas due to its outstanding ability. Also, we archived 1.5 GJ/t-CO_2 by SA-VSA (Fig. 4.12). The energy efficiency of a coal-fired power plant with a CO_2-capture system was estimated to improve by about 2% when the RITE solid sorbent was used instead of an advanced liquid amine solvent (2.5 GJ/t-CO_2).

R&Ds of solid sorbents are under progress also in other countries. However, conventional solid sorbents generally require high-temperature processes, which is not favorable in view of neither energy consumption nor degradation issue. On the other hand, the RITE solid sorbent exhibits the unique characteristic in regeneration capability at low temperature and at a low energy cost.

In 2015, we launched a new project by METI for practical application of CO_2-capture technologies (R&D of advanced solid sorbent for commercialization), in which bench-scale plant tests by the moving-bed system for coal-fired flue gas and simulation studies of the moving-bed system are underway in collaboration with Kawasaki Heavy Industries, Ltd.

Simultaneously, we are optimizing our solid sorbent toward practical application. RITE solid sorbent using the novel amines shows higher-performance than the solid sorbent using commercial amines. In addition, Improved RITE solid sorbent

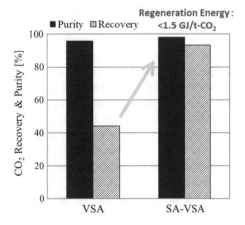

Fig. 4.12 CO_2-capture performance using RITE solid sorbent

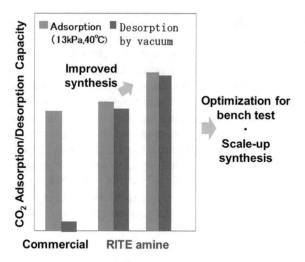

Fig. 4.13 Synthesis of RITE amine for bench test

obtained by an amine-synthesis method suitable for a large scale further enhanced the performance (Fig. 4.13).

Using the novel amines obtained by the improved-synthesis method for a large scale, we also prepared RITE solid sorbents in a larger scale and started the bench-scale plant test from November 2016.

4.5   Summary and Future Prospects

Absorption method has already been established as a commercial technology, it is suitable for large capacity because scale merit is obtained due to treatment in one line, but it requires a tall absorbing tower and regeneration tower and it takes time and labor to operate. On the other hand, in the case of small equipment, PSA is more compact and easier to use, which has the merit that a post-treatment system such as a trap of slip amine is unnecessary. Particularly, if a chemically stable material is used as the adsorbent, there is no risk of contamination by new substances of environmental concern in the exhaust gas or waste liquid treatment is unnecessary. From the viewpoint of technology for preserving the global environment, this is a great advantage compared to other separation techniques.

Several new sorbents with high CO$_2$ adsorption were developed. For mesoporous silica-based amine-modified materials, the pore size, pore volume and surface area of the supports, loaded amine, amine composition, and molecular structures of the blended amines in addition to the adsorption temperature were found to play important roles and greatly affect the CO$_2$-adsorption performance. Furthermore, detailed experimental investigations revealed the CO$_2$-adsorption mechanisms of these materials. Several problems must be resolved for actual applications. However, to establish commercially available adsorptive CO_2-separation techniques for large-point sources, we will have to accelerate the development and achieve energy and cost reductions (<1.5 GJ/ton-CO_2).

In addition to the large-scale CO_2 recovery from stationary sources, the CO_2-adsorptive separation technology may be applicable to removal of CO_2 in the closed space and direct CO_2 capture from air (direct air capture, DAC), for example. We are currently examining the applicability of these.

Acknowledgements This project is supported by the Ministry of Economy, Trade and Industry (METI), Japan, and the New Energy and Industrial Technology Development Organization (NEDO), Japan.

References

1. Breck DW (1974) Zeolite molecular sieves. Wiley
2. Chang ACC, Chuang SSC, Gray M, Soong Y (2003) In-situ infrared study of CO_2 adsorption on SBA-15 grafted with γ-(aminopropyl)triethoxysilane. Energy Fuels 17:468
3. Cheng Y, Kondo A, Noguchi H, Kajiro H, Urita K, Ohba T, Kaneko K, Kanoh H (2009) Reversible structural changes of Cu-MOF on exposure to water and its CO_2 adsorptivity. Langmuir 25:4510–4513
4. Choi S, Drese JH, Jones CW (2009) Adsorbent materials for carbon dioxide capture from large anthropogenic point source. ChemSusChem 2:796–854
5. Corma A (1997) From microporous to mesoporous molecular sieve materials and their use in catalysis. Chem Rev 97:2373
6. Dao DS, Yamada H, Yogo K (2013) Large-pore mesostructured silica impregnated with blended amines for CO_2 capture. Ind Eng Chem Res 52:13810–13817
7. Dasgupta S, Nanoti A, Gupta P, Jena D, Goswani AN, Garg MO (2009) Carbon dioxide removal with mesoporous adsorbents in a single column pressure swing adsorber. Sep Sci Technol 44:3973
8. Drage TC, Snape CE, Stevens LA, Wood J, Wang J, Cooper AI, Dawson R, Guo X, Satterley C, Irons R (2012) Materials challenges for the development of solid sorbents for post-combustion carbon capture. J Mater Chem 22:2815
9. Feng X, Fryxell GE, Wang L-Q, Kim AY, Liu J, Kemner KM (1997) Functionalized monolayers on ordered mesoporous supports. Science 276:923
10. Feng X, Hu G, Hu X, Xie G, Xie Y, Lu J, Luo M (2013) Tetraethylenepentamine-modified siliceous mesocellular foam (MCF) for CO_2 capture. Ind Eng Chem Res 52:4221
11. Fujiki J, Yogo K (2013) Polyethyleneimine-functionalized biomass-derived adsorbent beads for carbon dioxide capture at ambient conditions. Chem Lett 42:1484
12. Fujiki J, Chowdhury FA, Yamada H, Yogo K (2017) Highly efficient post-combustion CO_2 capture by low-temperature steam-aided vacuum swing adsorption using a novel polyamine-based solid sorbent. Chem Eng J 307:273–282
13. Herm ZR, Swisher JA, Smit B, Krishna BR, Long JR (2011) Metal-organic frameworks as adsorbents for hydrogen purification and precombustion carbon dioxide capture. J Am Chem Soc 133:5664–5667
14. Hiyoshi N, Yogo K, Yashima T (2004) Adsorption of carbon dioxide on modified SBA-15 in the presence of water vapor. Chem Lett 33:510
15. Hiyoshi N, Yogo K, Yashima T (2005) Adsorption of carbon dioxide on aminosilane-modified mesoporous silica. J Jpn Petrol Inst 48:29

16. Hiyoshi N, Yogo K, Yashima T (2005) Adsorption characteristics of carbon dioxide on organically functionalized SBA-15. Microporous Mesoporous Mater 84:357
17. Hiyoshi N, Yogo K, Yashima T (2008) Adsorption of carbon dioxide on amine-modified MSU-H silica in the presence of water vapor. Chem Lett 37:1266
18. Huang HY, Yang RT, Chinn D, Munson CL (2003) Amine-grafted MCM-48 and silica xerogel as superior sorbents for acidic gas removal from natural gas. Ind Eng Chem Res 42:2427
19. Inagaki S, Fukushima Y, Kuroda K (1993) Synthesis of highly ordered mesoporous materials from a layered polysilicate. Chem Soc Chem Commun 680–682
20. Inagaki S, Guan S, Fukusima Y, Ohsuna T, Terasaki O (1999) Novel mesoporous materials with a uniform distribution of organic groups and inorganic oxide in their frameworks. J Am Chem Soc 121:9611
21. Inui T, Okugawa Y, Yasuda M (1988) Relationship between properties of various zeolites and their carbon dioxide adsorption behaviors in pressure swing adsorption operation. Ind Eng Chem Res 27:1103
22. Kitagawa S, Matsuda R (2007) Chemistry of coordination space of porous coordination polymers. Coord Chem Rev 251:2490–2509
23. Kitagawa S, Kitaura R, Noro S (2004) Functional porous coordination polymers. Angew Chem Int Ed 43:2334–2375
24. Kizzie AC, Wong-Foy AG, Matzger AJ (2011) Effect of humidity on the performance of microporous coordination polymers as adsorbents for CO_2 capture. Langmuir 27:6368–6373
25. Knofel C, Descarpentries J, Benzoauia A, Zenlenal V, Mornet S, Llewellyn PL, Hornebecq V (2007) Functionalised micromesoporous silica for the adsorption of carbon dioxide. Microporous Mesoporous Mater 99:79
26. Kresge CT, Lenowicz ME, Ross WJ, Vartulli JC, Beck JS (1992) Ordered mesoporous molecular sieves synthesized by a liquid-crystal template mechanism. Nature 359:710–712
27. Kuppler RJ, Timmons DJ, Fang Q-R, Jli J-R, Makal TA, Young MD, Yuan D, Zhao D, Zhuang W, Zhou H-C (2009) Potential applications of metal-organic frameworks. Coord Chem Rev 253:3042–3066
28. Lee J-S, Kim J-H, Kim J-T, Suh J-K, Lee J-M, Lee C-H (2002) Adsorption equilibria of CO_2 on zeolite 13X and zeolite X/activated carbon composite. J Chem Eng Data 47:1237–1242
29. Liu W, Choi S, Drase JH, Hornbostel M, Krishman G, Eisenberger PM, Jones CW (2010) Steam-stripping for regeneration of supported amine-based CO_2 adsorbents. ChemSusChem 3:899–903
30. Miyamoto M, Takayama A, Uemiya S, Yogo K (2012) Study of gas adsorption properties of amideamine-loaded mesoporous silica for examing its use in CO_2 separation. J Chem Eng Jpn 45:395
31. Moellmer J, Moeller A, Driesbach F, Glaeser R, Staudt R (2011) High pressure adsorption of hydrogen, nitrogen, carbon dioxide and methane on the metal-organic framework HKUST-1. Microporous Mesoporous Mater 138:140–148
32. Quyen TV, Yamada H, Yogo K (2018) Exploring the role of imidazoles in amine-impregnated mesoporous silica for CO_2 capture. Ind Eng Chem Res 57:2638–2644
33. Rowsell JLC, Yaghi OM (2004) Metal-organic frameworks: a new class of porous materials. Microporous Mesoporous Mater 73:3–14
34. Saima H, Mogi Y, Haraoka T (2013) Development of PSA technology for the separation of carbon dioxide from blast furnace gas. JFE Tech Rep 32:44
35. Samanta A, Zhao A, George K, Shimazu H, Sarkar P, Gupta R (2012) Post-combustion CO_2 capture using solid sorbents: a review. Ind Eng Chem Res 51:1438
36. Schoenecker PM, Carson CG, Jasuja H, Flemming CJJ, Walton KS (2012) Effect of water adsorption on retention of structure and surface area of metal-organic frameworks. Ind Eng Chem Res 51:6513–6519
37. Sirwardane RV (2005) U.S. patent 6,908,497 B1
38. Son WJ, Choi JS, Ahn WS (2008) Adsorptive removal of carbon dioxide using polyethyleneimine-loaded mesoporous silica materials. Microporous Mesoporous Mater 113:31

39. Wang Q, Luo J, Zhong Z, Borgna A (2011) CO_2 capture by solid sorbents and their applications: current status and new trends. Energy Environ Sci 4:42
40. Wang X, Li H, Liu H, Hou X (2011) AS-synthesized mesoporous silica MSU-1 modified with tetraethylenepentamine for CO_2 adsorption. Microporous Mesoporous Mater 142:564
41. Wang J, Chen H, Zhou H, Liu X, Qiao W, Long D, Ling L (2013) Carbon dioxide capture using polyethyleneimine-loaded mesoporous carbons. J Environ Sci 25:124
42. Watabe T, Yogo K (2013) Isotherms and isosteric heats of adsorption for CO_2 in amine-functionalized mesoporous silicas. Sep Purif Technol 120:20
43. Wei J, Liao L, Xiao Y, Zhang P, Shi Y (2010) Capture of carbon dioxide by amine-impregnated as-synthesized MCM-41. J Environ Sci 22:1558
44. Xu X, Song C, Andresen JM, Miller BG, Scaroni AW (2002) Novel polyethylenimine-modified mesoporous molecular sieve of MCM-41 type as high-capacity adsorbent for CO_2 capture. Energy Fuels 16:1463
45. Xu X, Song C, Andresen JM, Miller BG, Scaroni AW (2003) Preparation and characterization of novel CO_2 "molecular basket" adsorbents based on polymer-modified mesoporous molecular sieve MCM-41. Microporous Mesoporous Mater 62:29
46. Yamada H, Fujiki J, Chowdhury FA, Yogo K (2018) Effect of isopropyl-substituent introduction into tetraethylenepentamine based solid sorbents for CO_2 capture. Fuel 214:14–19
47. Yan W, Tang J, Bian Z, Hu J, Liu H (2012) Carbon dioxide capture by amine-impregnated mesocellular-foam-containing template. Ind Eng Chem Res 51:3653
48. Yue MB, Sun LB, Cao Y, Wang ZJ, Wang Y, Yu Q, Zhu JH (2008) Promoting the CO_2 adsorption in the amine-containing SBA-15 by hydroxyl group. Microporous Mesoporous Mater 114:74
49. Zelenak V, Badanikova M, Halamova D, Cejka J, Zukal A, Murafa N, Goerigk G (2008) Amine-modified ordered mesoporous silica: effect of pore size on carbon dioxide capture. Chem Eng J 144:336
50. Zhang X, Zheng X, Zhang S, Zhao B, Wu W (2012) AM-TEPA impregnated disordered mesoporous silica as CO_2 capture adsorbent for balanced adsorption-desorption properties. Ind Eng Chem Res 51:15163

Chapter 5
Membrane for CO_2 Separation

Abstract Among various CO_2-capture technologies, membrane separation is considered as one of the promising solutions, because of its energy efficiency and operation simplicity. Many research and development are conducted for the (1) CO_2/N_2 (post-combustion: CO_2 separation from flue gas), (2) CO_2/CH_4 (CO_2 separation from natural gas), and (3) CO_2/H_2 (pre-combustion: CO_2 separation from integrated gasification combined cycle (IGCC) processes). In this chapter, research and development of various types of membranes (polymeric membranes, inorganic membranes, ionic liquid membranes, and facilitated transport membranes) for these applications are briefly reviewed. In the latter part of the chapter, the development of "molecular gate" membrane modules by the authors for the pre-combustion to separate CO_2 from H_2 at the integrated coal gasification combined cycle (IGCC) power plants are reviewed.

Keywords Membrane · Post-combustion · Pre-combustion · Molecular gate membrane

5.1 Membrane Separation Technology for CO_2 Separation

In general, the membrane separation method is the most promising in terms of both the operation and capital costs, because it allows energy saving and space saving for the CO_2-capture process. Various types of membranes were developed for CO_2 separation from various applications.

There are three major targets in membrane separation for CO_2 capture: (1) CO_2/N_2 (Post-combustion: CO_2 separation from flue gas), (2) CO_2/CH_4 (CO_2 separation from natural gas), and (3) CO_2/H_2 (pre-combustion: CO_2 separation from integrated gasification combined cycle (IGCC) processes).

In the case of post-combustion, CO_2 separation from flue gas, more than half of the cost of membrane separation goes toward powering the vacuum pump to evacuate the permeate side of the membrane. In addition, the costs of the membrane module and piping are high, because the pressure difference between the feed and the permeate side is low and a large membrane area is needed. In this case, high CO_2 permeability is more important than high selectivity to reduce the cost of the membrane modules.

© The Author(s), under exclusive license to Springer Nature Switzerland AG 2019   65
S. Nakao et al., *Advanced C O₂ Capture Technologies*,
SpringerBriefs in Energy, https://doi.org/10.1007/978-3-030-18858-0_5

On the other hand, in the case of pre-combustion, CO_2 separation in IGCC processes, a significant reduction in the CO_2-capture cost is expected via the use of membrane technology, because a vacuum pump or compressor is not needed for high-pressure gas separations. In this case, both CO_2 permeability and CO_2/H_2 selectivity are important to separate CO_2 effectively.

5.1.1 Polymeric Membrane

There are many studies of CO_2 selective polymer membranes for the separation of CO_2/CH_4 and CO_2/N_2 gas mixtures. On the other hand, there are comparatively few polymeric membranes that can be utilized for the selective recovery of CO_2 over H_2.

Polymeric membranes made from glassy polymers such as cellulose acetate and polyimide have exhibited practical use in selective CO_2 separation from CO_2/CH_4 gas mixtures. However, CO_2/CH_4 separation greatly decreases under high CO_2 partial pressures because of CO_2-induced plasticization. Koros et al. reported that cross-linked polyimide membranes showed enhanced resistance to CO_2 plasticization [23].

Poly(ethylene glycol) (PEG) has a high physical affinity toward CO_2 and was expected to be a viable CO_2-separation membrane material. However, pure PEG exhibited very low CO_2 permeability, owing to its crystallization. Freeman et al. developed cross-linked PEG membranes in order to prevent crystallization. The cross-linked PEG membranes exhibited favorable interactions with CO_2, which enhanced the solubility of CO_2 over that of H_2 and showed a CO_2/H_2 selectivity of about 10–35 °C and 25 at −20 °C [18]. Wessling et al. developed a PEG block copolymer membrane and measured the separation properties of the membrane and obtained a CO_2/H_2 selectivity of 10 at 35 °C [11]. Peinemann et al. developed a PEG block copolymer with CO_2 affinity and obtained a CO_2/H_2 selectivity of 10.8 at 30 °C [3].

Thermally rearranged polymers (TR polymers) are a novel polymer membrane in which the molecular sizes of the interchains are controlled by heat treatment. High CO_2/CH_4 separations by TR polymer membranes were maintained under high pressures because of its enhanced resistance to plasticization [24]. It was also reported that microporous organic polymer (MOP) membranes with high affinity to CO_2 displayed excellent CO_2 separation [5].

Mixed matrix membranes (MMM) have been developed by incorporating fillers (inorganic nanoparticles, etc.) into the polymeric matrix to improve separation performance. Shin et al. prepared PEG-MEA and graphene oxide additives in Pebax 1657 [27]. Zhu et al. prepared hollow fiber membranes with composite microcapsules made of polydopamine/poly(ethylene glycol) (PEG) [35]. Zhang et al. prepared MMM with aminosilane-functionalized graphene oxide nanosheets as additives [33]. Huang et al. prepared pebax/ionic liquid-modified graphene oxide MMM [10]. Sabetghadam et al. prepared thin MMM with a metal–organic framework (MOF) nanosheets [25]. Liu et al. prepared MMM with zeolite-like MOF nanocrystals [19].

CO_2 separation from flue gas using membranes is performed under low-pressure ratio between the feed side and permeate side of the membrane. The improvement in CO_2 permeability is important in terms of lowering the system cost and membrane area. Merkel et al. (Membrane Technology and Research (MTR)) proposed a new system to obtain a CO_2 partial pressure difference using air as a sweep gas at a low energy cost. In addition, a membrane module with high CO_2 permeability (Polaris™ membrane) was developed [20].

5.1.2 Inorganic Membrane

As for inorganic membranes, zeolite membranes and carbon membranes among others have been reported for CO_2 separation. Inorganic membranes have appropriately sized pores that can act as molecular sieves to separate gas molecules by effective size. In addition, inorganic membranes with strong CO_2 affinities show high CO_2 selectivity over N_2 and CH_4.

Noble et al. reported that SAPO-34 showed a high CO_2/CH_4-separation performance [34]. Bae et al. reported that organic/inorganic composite membranes prepared by incorporating MOF (ZIF-90) into a polymeric matrix also showed a high-CO_2/CH_4-separation performance [1]. Wang et al. prepared all-silica DDR zeolite membranes for CO_2/CH_4 separation by microwave-aided hydrothermal synthesis [31]. Dong et al. reported the thin ceramic-carbonate dual-phase membranes for CO_2 separation from syngas [4]. Shin et al. prepared carbon molecular sieve hollow fiber membranes by using compatible polymer/polysilsesquioxane blends as precursors [28].

5.1.3 Ionic Liquid Membrane

Ionic liquid membranes have received increasing interest and have been studied in recent years because of their low vapor pressures and stability at high temperatures. Polymerized ionic liquid membranes were prepared for the separation of CO_2/N_2, CO_2/CH_4, etc., by Noble et al. [2]. Amino-containing ionic liquids were investigated in the separation of CO_2/H_2 by Myers et al. [21] and a CO_2/H_2 selectivity of 15 was obtained at 85 °C. Matsuyama et al. reported amino-containing ionic liquid membranes for the separation of CO_2/CH_4 and the membrane showed constant separation abilities for 260 days $(\alpha_{CO_2/CH_4} = ca.60)$ [9]. Kasahara et al. prepared facilitated transport membranes containing amine-functionalized task-specific ionic liquids [15]. Nikolaeva et al. reported cellulose-based poly-ionic membranes for CO_2/N_2 and CO_2/CH_4 separation [22].

5.1.4 Facilitated Transport Membrane

Facilitated transport membranes for CO_2 separation were prepared by impregnating pores of microporous support membranes or polymer matrices with carrier solutions such as amines and alkali metal carbonates, which have a chemical affinity to CO_2. CO_2 carrier incorporated membrane can react selectively and reversibly with CO_2. The CO_2 transport membrane rate can be facilitated because CO_2 carrier of the reaction product can transport through the membrane, in addition to CO_2 transport membrane of the physical solution–diffusion mechanism. On the other hand, other gases, such as N_2, H_2, CO, transport membrane only by solution–diffusion mechanism. As a result, the CO_2 selectivity of facilitated transport membranes can be extremely high at low CO_2 partial pressures.

Ho et al. developed facilitated transport membranes by blending amines with poly(vinyl alcohol) (PVA) [36]. These membranes showed a CO_2/H_2 selectivity of 300 at 110 °C and 100 at 150 °C. Matsuyama et al. reported facilitated transport membranes prepared by the immobilization of 2,3-diaminopropionic acid and cesium carbonate in a PVA/poly(acrylic acid) copolymer matrix and the resulting membrane showed a CO_2/H_2 selectivity of 432 at 160 °C [32]. Hägg et al. developed a CO_2/N_2-separation membrane by blending PVA and poly(vinylamine) (PVAm). The composite membrane with a selective layer thickness of 0.3 μm was prepared by casting a solution of PVA/PVAm on a polysulfone (PSf) substrate membrane [26]. Vakharia et al. conducted scale-up fabrication of thin facilitated transport membranes by roll-to-roll coating techniques [30]. Han et al. prepared carbon nanotube-reinforced facilitated transport membranes [8].

5.2   Development of "Molecular Gate" Membrane

The government of Japan declared a goal to reduce CO_2 emissions by half of those measured in 2005 as part of the "Cool Earth 50" project. One promising means to reduce CO_2 emission is the development of integrated coal gasification combined cycles with CO_2 capture and storage (IGCC-CCS) (Fig. 5.1).

In the IGCC-CCS, CO_2-separation membranes will play an important role in reducing CO_2-capture costs. Estimates indicate that the CO_2-capture cost from a pressurized gas stream using a membrane can be 1500 JPY/t-CO_2 or less, provided that membranes with excellent CO_2/H_2 separation abilities are developed.

Sirkar et al. reported excellent CO_2/N_2 selectivity using a viscous and nonvolatile poly(amidoamine) (PAMAM) dendrimer as an immobilized liquid membrane under isobaric and saturated water vapor test conditions [17]. Following this report, we developed a CO_2 molecular gate membrane, with the goal of producing a novel, high-performance separation membrane. Figure 5.2 shows the schematic illustration of the working principles of the molecular gate membrane.

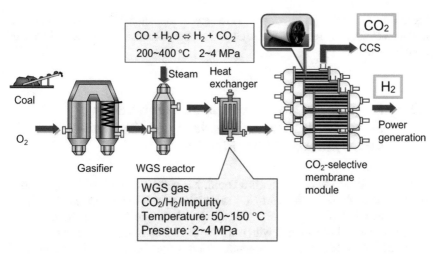

Fig. 5.1 Schematic of the IGCC process with CO_2 capture by CO_2 selective membrane modules

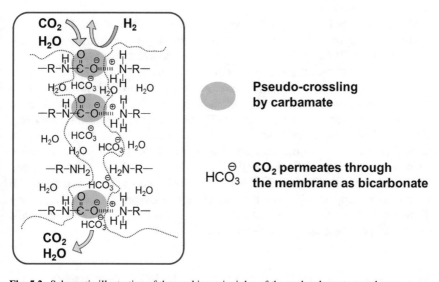

Fig. 5.2 Schematic illustration of the working principles of the molecular gate membrane

Under humidified conditions, CO_2 reacts with amino groups in the membrane to form either carbamate or bicarbonate, which then block the passage of H_2. Consequently, the amount of H_2 diffusing to the other side of the membrane is greatly reduced and high concentrations of CO_2 can be obtained. Therefore, a high CO_2/H_2 selectivity is obtained.

In this section, research regarding molecular gate membranes for investigating post-combustion and pre-combustion is presented. In Sect. 5.2.1, the development of the membranes by in situ coating for CO_2 removal from flue gas (i.e., post-combustion and CO_2/N_2 separation) is reviewed. In Sect. 5.2.2, the development of membranes for CO_2 removal from IGCC (i.e., pre-combustion and CO_2/H_2 separation) is reviewed.

5.2.1   Development of Commercial-Sized Dendrimer Composite Membrane Modules by In Situ Coating for CO_2 Removal from Flue Gas

Because the PAMAM dendrimer is a liquid, it is necessary to affix it to appropriate support materials. The excellent CO_2 selectivity of this dendrimer can be applied in the separation of CO_2 from fossil fuel emissions, provided that the dendrimer can be successfully fabricated with a stable membrane configuration and sufficient tolerance for practical use (e.g., pressure differences). Toward this end, a composite membrane would be considered an appropriate configuration for preparing stable, selective dendrimer layers.

Composite membranes typically possess a selective layer affixed to the top of mesoporous support, where the selective layer and support are fabricated from different materials. The selective layer on the composite membrane must be as thin as possible in order to maintain economical fluxes. Composite membranes offer a more flexible approach because of their cost-effectiveness, ease of fabrication, and the many combinations in which various supports and selective layers can be prepared.

At RITE, we successfully prepared PAMAM dendrimer composite membranes using chitosan as the gutter layer [6, 13, 16] and demonstrated the possibility of fabricating solid stable membranes from liquid viscous PAMAM dendrimers. Either ethylene glycol diglycidyl ether (EGDGE) or glutaraldehyde (GA) was used as a cross-linker, and ultrafiltration (UF) hollow fiber membranes constructed from PSF were used as support substrates for composite membranes. The outer and inner diameters of the hollow fiber membranes were 1900–1100 μm, respectively, while the molecular weight cutoff of the membrane was 6000.

5.2.1.1   Preparation of Composite Membrane by In Situ Modification

We developed a new method to coat PAMAM dendrimer layers on the support substrate (entire module) called in situ modification (IM). Figure 5.3 shows a schematic diagram of IM. IM is utilized for the preparation of ultrathin functional layers of membrane materials and is applicable with modules of any size.

A solution containing the membrane materials is directly circulated in the lumen of the hollow fiber membranes within the membrane module, while the shell of the hollow fiber membranes is evacuated. As shown in Fig. 5.3c, if a hydrophobic porous substrate and hydrophilic solution are applied, the hydrophilic solution cannot penetrate the hydrophobic porous substrate, but instead generates a gas–liquid interface

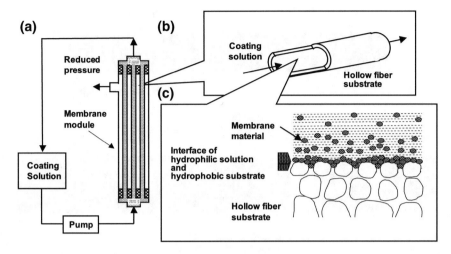

Fig. 5.3 Schematic diagram of the IM method

at the substrate surface. Because of the reduced pressure on the opposite side of the membrane, the solvent evaporates from the interface, resulting in a high solute concentration in the vicinity of the interface. Finally, the membrane material precipitates to form an ultrathin layer on or beneath the surface. One of the main advantages of IM is that the membrane pores at the surface are covered with molecules smaller than the pore size. In this study, a PSF-UF membrane and an aqueous chitosan solution were selected as the hydrophobic and hydrophilic media, respectively. Even though the average pore size of PSF is far larger than those of chitosan, the interfacial precipitation is likely to create an effective uniform chitosan layer.

We began with relatively small membrane modules with three hollow fiber membranes in the module housing (length = 20 cm and diameter = 3/8 in.), such that the effective membrane area of the module was about 18 cm^2.

After the development of a small module (pencil module and length: 200 mm), the development of commercial-sized membrane modules for CO_2 separation of ambient flue gas by IM was developed.

Figure 5.4 shows the photographs of the membrane modules: (a) pencil modules with lengths of 200 mm (three hollow fibers in a module) and 800 mm (seven hollow fibers in a module) and (b) commercial-sized modules.

In the gas permeation experiment, the apparatus was housed within a rectangular, open metal frame support. The modules were uninsulated with no heat tracing and therefore experiments were conducted at ambient temperature (ca. 25 °C), because the module was too large to fit in the air oven.

CO_2/N_2-separation performances of five commercial-sized membrane modules are shown in Figs. 5.6, 5.7, 5.8, 5.9, and 5.10. The CO_2 permeance and CO_2/N_2 selectivity ranged from 1.5 to 2.2 × 10^{-7} m^3 (STP) m^{-2} s^{-1} kPa^{-1} (average: 1.7 × 10^{-7} m^3 (STP) m^{-2} s^{-1} kPa^{-1}, standard deviation: ±0.3 × 10^{-7} m^3 (STP) m^{-2}

(a) Pencil modules
200, 800 mm in length, 3/8 inch in diameter

(b) commercial-sized modules.
1100 mm in length, 1 inch in diameter

Fig. 5.4 Photographs of the membrane modules. **a** Pencil modules and **b** commercial-sized modules

s^{-1} kPa^{-1}) and 110–170 (average: 150, standard deviation: ±20), respectively. All membrane modules had a separation factor of more than 100, indicating a superior performance than those of many reported polymeric membranes.

A pencil module (length: 200 mm and membrane area: 17 cm^2) had a CO$_2$/N$_2$ selectivity of 400 and Q_{CO_2} value of 1.6 × 10^{-7} m^3 (STP) m^{-2} s^{-1} kPa^{-1} at 40 °C [16]. Therefore, the commercial-sized membrane module did not perform as well as the pencil module with respect to the CO$_2$-separation factor. Potentially, this was because of: (1) the lower operating temperature (ca. 25 °C in this work compared with 40 °C in the previous work [18]), (2) the difference in the water vapor pressure on the permeate side, and (3) the difference in the cross-linker (GA vs. EGDGE).

In summary, IM was successfully applied not only to pencil modules but also to large, commercial-sized membrane modules with a separation factor of more than 100 under practical operating conditions (i.e., using the pressure difference method).

5.2.2 Development of Poly(amidoamine) Dendrimer/Polymer Hybrid Membrane Modules for CO$_2$ Removal from IGCC

In this section, PAMAM dendrimer/polymer hybrid membranes using either poly(ethylene glycol) (PEG) or poly(vinyl alcohol) as the polymer matrix are reviewed. The PAMAM dendrimer/cross-linked polymer hybrid membrane shows great potential for CO_2 separation from H_2 in high-pressure applications such as IGCC.

5.2.2.1 Poly(amidoamine) Dendrimer/Poly(ethylene glycol) Hybrid Membranes

To improve CO_2 separation under pressure difference conditions, hybrid membranes were developed by incorporating or immobilizing PAMAM dendrimer into a polymer matrix. We have succeeded in immobilizing the dendrimer in a cross-linked poly(ethylene glycol) (PEG) obtained by photopolymerization of PEG dimethacrylate (PEGDMA) [29]. PAMAM/PEGDMA hybrid membrane containing 50 wt% PAMAM dendrimer showed high CO_2-separation properties over H_2, for example, $\alpha_{CO_2/H_4} > 500$ at 5 kPa CO_2 partial pressure and 80% relative humidity under ambient conditions.

We also developed hybrid membranes composed of PAMAM dendrimer and PEGDMA and other cross-linkers (such as trimethylolpropane trimethacrylate (TMPTMA)) for high-pressure applications (Fig. 5.5).

The composite membranes, composed of the PAMAM dendrimer/PEG hybrid thin layer on the substrate membrane, were successfully prepared. The resulting PAMAM dendrimer/PEG hybrid membrane exhibited an excellent CO_2/H_2 selectivity of 39 with CO_2 permeance of 1.1×10^{-10} m^3 (STP)/(m^2 s Pa) at 560 kPa CO_2 partial under 700 kPa feed pressure with 80% relative humidity at 40 °C.

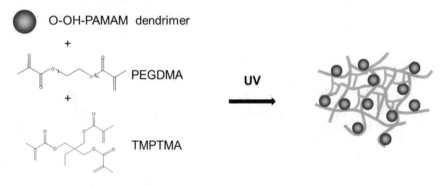

Fig. 5.5 Scheme of immobilization of the dendrimer in a cross-linked poly(ethylene glycol) (PEG) obtained by photopolymerization of PEG dimethacrylate (PEGDMA) and trimethylolpropane trimethacrylate (TMPTMA)

5.2.2.2   Poly(amidoamine) Dendrimer/Poly(vinyl alcohol) Hybrid Membranes

In the previous section, PAMAM dendrimers immobilized in cross-linked PEG by photopolymerization of PEG dimethacrylate in the presence of the dendrimer were described. In this section, the utilization of PVA as another polymeric matrix, in which CO$_2$ carriers were incorporated, will be described. The PVA matrix is compatible with the dendrimer because the hydroxyl group of PVA has an affinity toward the primary amine of the dendrimer via hydrogen bonding. PVA was cross-linked with an organic metal, di-isopropoxy-bis(triethanol aminato) titanium (Ti cross-linker) to enhance the mechanical properties for use under elevated process pressures.

Figure 5.6 shows a schematic diagram of the cross-linking of PVA in the presence of the PAMAM dendrimer [7].

PVA cross-linking was carried out with a Ti cross-linker in an aqueous solution and the dendrimer was incorporated into the cross-linked PVA matrix to form a self-standing membrane. The cross-linker was chosen for the following reasons:

(1) The Ti cross-linker is highly selective and only reacts with the hydroxyl group of PVA in aqueous media.

(2) The cross-linking reaction proceeds rapidly and effectively under mild conditions, minimizing unfavorable side reactions.

(3) The by-product (isopropanol) is readily eliminated by evacuation from the resulting polymeric membrane.

The dendrimer content and the membrane thickness were readily controlled by this robust preparation method, and subsequent leakage of the liquid dendrimer was negligible. Because of the high selectivity of the cross-linker toward the hydroxyl group of PVA, the primary amine of PAMAM dendrimer was not involved in the cross-linking reaction and the dendrimer remained intact during membrane preparation.

The CO$_2$-separation properties as a function of CO$_2$ partial pressure were studied with a polymeric membrane containing PAMAM dendrimer at 41.6 wt%. The thickness of the membrane was 400 μm and the gas separation test was performed at 40 °C. The Q_{CO_2} decreased sharply with increasing CO$_2$ partial pressures from 5 to 80 kPa and reached a constant level above 80 kPa (2×10^{-12} m^3 (STP)/(m^2 s Pa)). However, Q_{H_2} did not change significantly at the CO$_2$ partial pressure of 5 ×

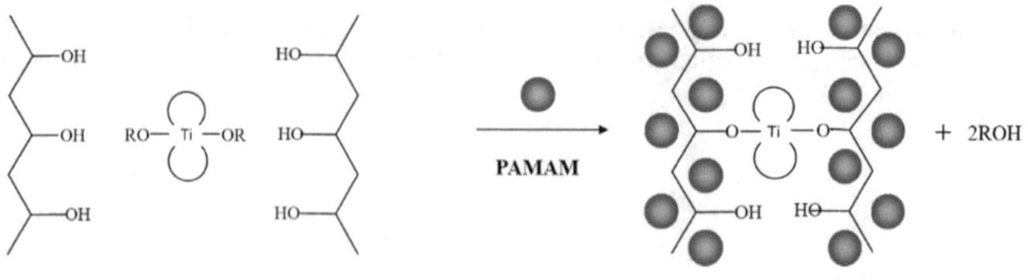

Fig. 5.6 Schematic diagram of a Ti cross-linked PVA matrix with PAMAM immobilized within it

10^{-14} m^3 (STP)/(m^2 s Pa). The PAMAM/PVA hybrid membrane retained a higher CO_2 selectivity under such CO_2 partial pressure, exhibiting a CO_2/H_2 selectivity of 32, which is much higher than the conventional polymeric membranes.

5.3 R&D Project for CO_2 Removal from IGCC Plant

Based on the achievements of molecular gate membranes by RITE, Molecular Gate Membrane module Technology Research Association (MGMTRA; Consists of Research Institute of Innovative Technology for the Earth (RITE) and private company) is developing molecular gate membranes and membrane elements for CO_2 separation with low energy and cost during the IGCC process [7, 12, 14].

In the previous project by Ministry of Economy, Trade and Industry (METI), Japan, "CO_2 Separation Membrane Module Research and Development Project" (2011FY-2014FY), we developed the molecular gate membranes that show high CO_2-separation performance under high-pressure conditions of 2.4 MPa (supposed pressure in IGCC process) using membranes prepared in the laboratory.

In the current project by METI and New Energy and Industrial Technology Development Organization (NEDO), Japan, "CO_2 Separation Membrane Module Practical Research and Development Project" (2015FY-), we are developing the membranes with large membrane area by continuous membrane-forming method, and also developing the membrane elements.

5.3.1 Membrane Preparation

We chose poly(vinyl alcohol) (PVA)-based materials as major component materials for pressure-durable polymeric matrix. Example materials and reaction scheme of PVA matrix are shown in Fig. 5.7.

As shown in Fig. 5.7, dendrimer was immobilized in cross-linked PVA matrix. Thin dendrimer/PVA hybrid membrane was formed on the porous support to achieve high CO_2 permeance.

5.3.2 Gas Permeation Performances

In the gas permeation experiments, helium (He) gas was used as alternative gas to H_2, for safety issue (molecular size of He is similar to that of H_2). Typical operating conditions of gas permeation experiment using simulated gas are as follows: CO_2/He gas mixture (CO_2/He = 40/60 vol%) was humidified at 50–80% relative humidity and then fed to a flat-sheet membrane cell or a membrane module. The total pressure at the feed side was 2.4 MPa. The total pressure at the permeate side was atmospheric

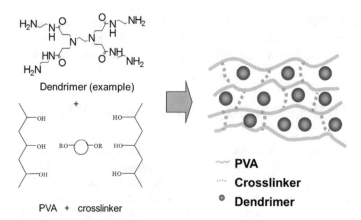

Fig. 5.7 Schematic of dendrimer/PVA hybrid membrane (an example)

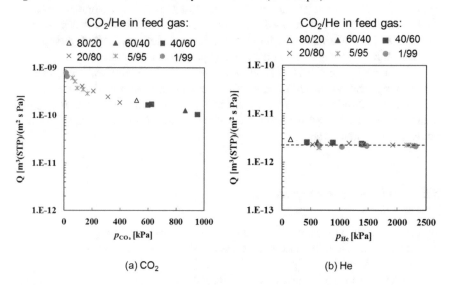

(a) CO₂   (b) He

Fig. 5.8 Effect of CO_2 partial pressure on permeance (Q). Operating conditions: temp: 85 °C, feed gas: 0.7–2.4 MPa; permeate gas: atmospheric pressure (Ar sweep gas)

pressure. The operating temperature was 85 °C. The CO_2 and He concentrations in both feed (retentate) and permeate gas were measured by gas chromatography.

Effect of CO_2 partial pressure on permeance (Q) using membrane1 (1.2 cm²) is shown in Fig. 5.8.

As shown in Fig. 5.8, CO_2 permeance decreased as CO_2 partial pressure increases, whereas He permeance was constant, not depending on CO_2 partial pressure. Interestingly, CO_2 permeation behavior can be summarized by CO_2 partial pressure, not total pressure, in the feed gas.

Fig. 5.9 Effect of CO₂ partial pressure on CO₂/He selectivity ($\alpha_{CO_2/He}$). Operating conditions: temp: 85 °C, feed gas: 0.7–2.4 MPa, permeate gas: atmospheric pressure (Ar sweep gas)

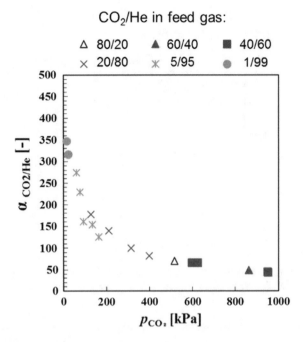

Effect of CO₂ partial pressure on CO₂/He selectivity (α) using membrane1 (1.2 cm²) is shown in Fig. 5.9.

As shown in Fig. 5.9, CO₂/He selectivity is also a function of CO₂ partial pressure. This result is due to the dependence of CO₂ permeance on CO₂ partial pressure.

In general, both permeance and selectivity are constants and not dependent on partial pressure, and the dependence of CO₂ permeance of this membrane is unique. It is considered that CO₂ permeation mechanism is not solution–diffusion mechanism, but that the reaction from CO₂ to carbamate or bicarbonate ion at feed side and the reverse reaction at the permeation side should be taken into account. On the other hand, He-permeation behavior can be explained by solution–diffusion mechanism.

Separation performances of the flat-sheet membranes and the 2 in. spiral membrane module are summarized in Fig. 5.10.

As shown in Fig. 5.10, similar dependence of CO₂-separation performance on CO₂ partial pressure was obtained for all the membranes (both membrane area of 1.2–58 cm²) and 2 in. membrane modules. Based on our cost simulation, one of the membranes (membrane2) showed the separation performance to achieve the project cost target (i.e., 1500 JPY/t-CO₂).

Comparison of CO₂/He and CO₂/N₂ separation performance is shown in Fig. 5.11.

CO₂ permeance was almost the same, not depending on CO₂/He or CO₂/N₂. On the other hand, N₂, which is larger molecular than He, showed lower permeance. As a result, CO₂/N₂ selectivity was an order of magnitude higher than CO₂/He selectivity. Therefore, it was found that N₂ does not affect the CO₂ separation performance.

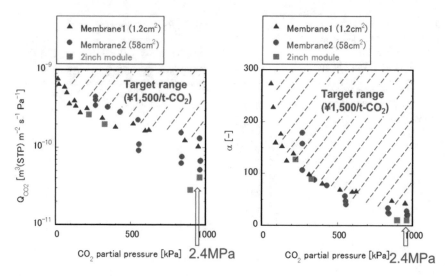

Fig. 5.10 Key results of CO₂ molecular gate membranes and spiral membrane modules Operating conditions: temp: 85 °C, feed gas: 0.7−2.4 MPa, permeate gas: atmospheric pressure (Ar sweep gas)

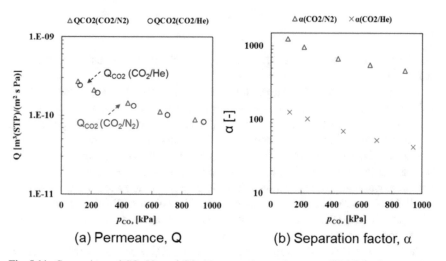

Fig. 5.11 Comparison of CO₂/He and CO₂/N₂ separation performance (85 °C, feed gas composition: CO₂/He or CO₂/N₂ = 40/60–5/95%, feed gas pressure: 2.4 MPa, feed gas humidity: 60% RH, permeate gas pressure: atmospheric pressure (Ar sweep))

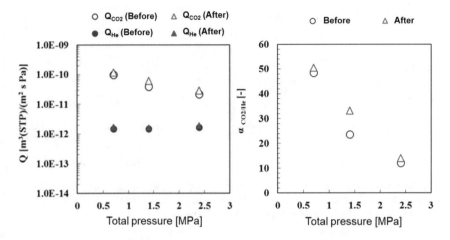

Fig. 5.12 Separation performance of the membrane before and after the H$_2$S exposure test. "Before": Before H$_2$S exposure test, "After": After H$_2$S exposure test. H$_2$S exposure test conditions: pressure: 2.4 MPa; temperature: 85 °C; gas composition: CO$_2$ (33%) + H$_2$S (500 ppm + N$_2$ balance (Relative humidity: ca. 80% RH). Exposure period: 7 d. Operating conditions: Temperature: 85 °C, feed gas composition: CO$_2$/He = 40/60%, feed gas pressure: 2.4 MPa, relative humidity in feed: 60% RH, permeate gas pressure: atmospheric pressure

The real gas from the coal gasifier contains trace amounts of impurities, such as CO, CH$_4$, H$_2$S, and COS. Among them, negative effect of H$_2$S on the membrane separation performance was concerned. So, the effect of H$_2$S exposure on the CO$_2$-separation performance was evaluated. The results are shown in Fig. 5.12.

As shown in Fig. 5.12, there was no significant effect of H$_2$S exposure on CO$_2$-separation performance. Therefore, it is suggested that the prepared membrane is resistant to H$_2$S.

Long-term stability test was conducted for more than 600 h at 2.4 MPa in the standard operating conditions without H$_2$S. The result is shown in Fig. 5.13.

As shown in Fig. 5.13, the membrane showed stable separation performance for at least 600 h at 2.4 MPa.

5.3.3 Fabrication of Membrane Elements

So far, membranes were prepared by single sheet coating method (Fig. 5.14a). This method is convenient for the screening of membrane materials in small scale. On the other hand, in the current project, the continuous membrane-forming method is conducted to produce the membranes continuously for the mass production of the flat-sheet membranes to fabricate the membrane elements (Fig. 5.14b).

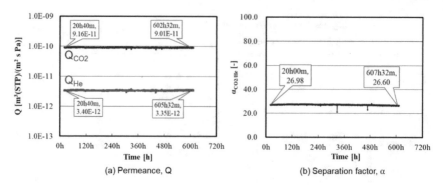

Fig. 5.13 Long-term stability of the membrane separation performance. Operating conditions: temperature: 85 °C, feed gas composition: $CO_2/He = 40/60\%$, feed gas pressure: 2.4 MPa, relative humidity in feed: 60% RH, permeate gas pressure: atmospheric pressure

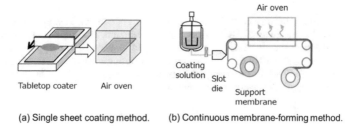

(a) Single sheet coating method.   (b) Continuous membrane-forming method.

Fig. 5.14 Schematic of **a** single sheet coating method and **b** continuous membrane-forming method

Photographs of CO_2 selective membranes, membrane elements, and membrane modules developed by MGMTRA are shown in Fig. 5.15. Membrane element is the structure with a large membrane area composed of the membrane, support membrane, spacer, etc. Membrane module is the structure in which the membrane element is placed.

5.4   Summary

As reviewed in this chapter, many research and development of novel CO_2 selective membranes are conducted for the CO_2/N_2, CO_2/CH_4 and CO_2/H_2 separation. It is expected that membrane process will be commercialized in the near future.

As for the development of molecular gate membrane modules, we have developed excellent membrane materials that show high-CO_2/H_2-separation performance under high pressure for IGCC process. In addition, durability of the membrane for H_2S and long-term stability were confirmed. We are developing the membranes with large membrane areas by continuous membrane-forming method and also developing the membrane elements for the mass production of membranes and membrane elements.

Fig. 5.15 Photographs of
CO_2 selective membrane,
membrane element, and
membrane module

CO_2 selective membrane

Membrane element
（4inch; L = 200mm）

Membrane module

Image of membrane element in
membrane module

Acknowledgements Development of CO_2 molecular gate membranes has been supported by the
Ministry of Economy, Trade and Industry (METI), Japan since 2003, and by the New Energy and
Industrial Technology Development Organization (NEDO), Japan since 2018.

References

1. Bae T-H, Lee JS, Qiu W, Koros WJ, Jones CW, Nair S (2010) A High-performance
   gas-separation membrane containing submicrometer-sized metal-organic framework crystal.
   Angew Chem Int Ed 49:9863–9866
2. Bara JE, Lessmann S, Gabriel CJ, Hatakeyama ES, Noble RD, Gin DL (2007) Synthesis and
   performance of polymerizable room-temperature ionic liquids as gas separation membranes.
   Ind Eng Chem Res 46:5397–5404
3. Car A, Stropnik C, Yave W, Peinemann K-V (2008) PEG modified poly(amide-b-ethylene
   oxide) membranes for CO2 separation. J Membr Sci 307:88–95
4. Dong X, Wu HC, Lin YS (2018) CO_2 permeation through asymmetric thin tubular ceramic-
   carbonate dual-phase membranes. J Membr Sci 564:73–81
5. Du N, Park HB, Robertson GP, Dal-Cin MM, Visser T, Scoles L, Guiver MD (2011) Polymer
   nano sieve membranes for CO_2-capture applications. Nat Mater 10:372–375
6. Duan S, Kouketsu T, Kazama S, Yamada K (2006) Development of PAMAM dendrimer composite membranes for CO_2 separation. J Membr Sci 283:2–6
7. Duan S, Taniguchi I, Kai T, Kazama S (2012) Poly(amidoamine) dendrimer/poly(vinyl alcohol)
   hybrid membranes for CO_2 capture. J Membr Sci 423–424:107–112
8. Han Y, Wu D, Ho WSW (2018) Nanotube-reinforced facilitated transport membrane for
   CO_2/N_2 separation with vacuum operation. J Membr Sci 567:261–271
9. Hanioka S, Maruyama T, Sotani T, Teramoto M, Matsuyama H, Nakashima K, Hanaki M,
   Kubota F, Goto M (2008) CO_2 separation facilitated by task-specific ionic liquids using a
   supported liquid membrane. J Membr Sci 314:1–4
10. Huang G, Isfahani AP, Muchtar A, Sakurai K, Shrestha BB, Qin D, Yamaguchi D, Sivaniah
    E, Ghalei B (2018) Pebax/ionic liquid modified graphene oxide mixed matrix membranes for
    enhanced CO_2 capture. J Membr Sci 565:370–379

11. Husken D, Visser T, Wessling M, Gaymans RJ (2010) CO_2 permeation properties of poly(ethylene oxide)-based segmented block copolymers. J Membr Sci 346:194–201
12. Kai T, Duan S, Ito F, Mikami S, Sato Y, Nakao S (2017) Development of CO_2 Molecular Gate Membranes for IGCC Process with CO_2 Capture. Energy Procedia 114:613–620
13. Kai T, Kouketsu T, Duan S, Kazama S, Yamada K (2008) Development of commercial-sized dendrimer composite membrane modules for CO_2 removal from flue gas. Sep Purif Tech 63:524–530
14. Kai T, Taniguchi I, Duan S, Chowdhury FA, Saito T, Yamazaki K, Ikeda K, Ohara T, Asano S, Kazama S (2013) Molecular gate membrane: Poly(amidoamine) dendrimer/polymer hybrid membrane modules for CO_2 capture. Energy Procedia 37:961–968
15. Kasahara S, Kamio E, Otani A, Matsuyama H (2014) Fundamental investigation of the factors controlling the CO_2 permeability of facilitated transport membranes containing Amine-functionalized task-specific ionic liquids. Ind Eng Chem Res 53:2422–2431
16. Kouketsu T, Duan S, Kai T, Kazama S, Yamada K (2007) PAMAM dendrimer composite membrane for CO_2 separation: Formation of a chitosan gutter layer. J Membr Sci 287:51–59
17. Kovvali AS, Chen H, Sirkar KK (2000) Dendrimer membranes: A CO_2-selective molecular gate. J Am Chem Soc 122(31):7594–7595
18. Lin H, Wagner EV, Freeman BD, Toy LG, Gupta RP (2006) Plasticization-enhanced hydrogen purification using polymeric membranes. Sci 311:639–642
19. Liu G, Labreche Y, Chernikova V, Shekhah O, Zhang C, Belmabkhout Y, Eddaoudi M, Koros WJ (2018) Zeolite-like MOF nanocrystals incorporated 6FDA-polyimide mixed-matrix membranes for CO_2/CH_4 separation. J Membr Sci 565:186–193
20. Merkel TC, Lin H, Wei X, Baker R (2010) Power plant post-combustion carbon dioxide capture: An opportunity for membranes. J Membr Sci 359:126–139
21. Myers C, Pennline H, Luebke D, Ilconich J, Dixon JK, Maginn EJ, Brennecke JF (2008) High temperature separation of carbon dioxide/hydrogen mixtures using facilitated supported ionic liquid membranes. J Membr Sci 322:28–31
22. Nikolaeva D, Azcune I, Tanczyk M, Warmuzinski K, Jaschik M, Sandru M, Dahl PI, Genua A, Lois S, Sheridan E, Fuoco A, Vankelecom IFJ (2018) The performance of affordable and stable cellulose-based poly-ionic membranes in CO_2/N_2 and CO_2/CH_4 gas separation. J Membr Sci 564:552–561
23. Omole IC, Adams RT, Miller SJ, Koros WJ (2010) Effects of CO_2 on a high performance hollow-fiber membrane for natural gas purification. Ind Eng Chem Res 49:4887–4896
24. Park HB, Jung CH, Lee YM, Hill AJ, Pas SJ, Mudie ST, Van Wagner E, Freeman BD, Cookson DJ (2007) Polymers with cavities tuned for fast selective transport of small molecules and ions. Sci 318:254–258
25. Sabetghadam A, Liu X, Gottmer S, Chu L, Gascon J, Kapteijn F (2019) Thin mixed matrix and dual layer membranes containing metal-organic framework nanosheets and Polyactive™ for CO_2 capture. J Membr Sci 570–571:226–235
26. Sandru M, Haukebo SH, Hagg M-B (2010) Composite hollow fiber membranes for CO_2 capture. J Membr Sci 346:172–186
27. Shin JH, Lee SK, Cho YH, Park HB (2019) Effect of PEG-MEA and graphene oxide additives on the performance of Pebax®1657 mixed matrix membranes for CO_2 separation. J Membr Sci 572:300–308
28. Shin JH, Yu HJ, An H, Lee AS, Hwang SS, Lee SY, Lee JS (2019) Rigid double-stranded siloxane-induced high-flux carbon molecular sieve hollow fiber membranes for CO_2/CH_4 separation. J Membr Sci 570–571:504–512
29. Taniguchi I, Duan S, Kazama S, Fujioka Y (2008) Facile fabrication of a novel high performance CO_2 separation membrane: Immobilization of poly(amidoamine) dendrimers in poly(ethylene glycol) networks. J Membr Sci 322:277–280
30. Vakharia V, Salim W, Wu D, Han Y, Chen Y, Lin Z, Ho WSW (2018) Scale-up of amine-containing thin-film composite membranes for CO_2 capture from flue gas. J Membr Sci 555:379–387

31. Wang M, Bai L, Li M, Gao L, Wang M, Rao P, Zhang Y (2019) Ultrafast synthesis of thin all-silica DDR zeolite membranes by microwave heating. J Membr Sci 572:567–579
32. Yegani R, Hirozawa H, Teramoto M, Himei H, Okada O, Takigawa T, Ohmura N, Matsumiya N, Matsuyama H (2007) Selective separation of CO_2 by using novel facilitated transport membrane at elevated temperatures and pressures. J Membr Sci 291:157–164
33. Zhang J, Xin Q, Li X, Yun M, Xu R, Wang S, Li Y, Lin L, Ding X, Ye H, Zhang Y (2019) Mixed matrix membranes comprising aminosilane-functionalized graphene oxide for enhanced CO_2 separation. J Membr Sci 570–571:343–354
34. Zhang Y, Tokay B, Funke HH, Falconer JL, Noble RD (2010) Template removal from SAPO-34 crystals and membranes. J Membr Sci 363:29–35
35. Zhu B, Liu J, Wang S, Wang J, Liu M, Yan Z, Shi F, Li J, Li Y (2019) Mixed matrix membranes containing well-designed composite microcapsules for CO_2 separation. J Membr Sci 572:650–657
36. Zou J, Ho WSW (2006) CO_2-selective polymeric membranes containing amines in crosslinked poly(vinyl alcohol). J Membr Sci 286:310–321

Printed in the United States
By Bookmasters